W0051188

STRUCTURE AND BONDING

Volume 26

Editors: J. D. Dunitz, Zürich
P. Hemmerich, Konstanz · R. H. Holm, Stanford
J. A. Ibers, Evanston · C. K. Jørgensen, Genève
J. B. Neilands, Berkeley · D. Reinen, Marburg
R. J. P. Williams, Oxford

With 15 Figures and 52 Tables

Springer-Verlag
Berlin Heidelberg GmbH 1976

ISBN 978-3-662-15843-2   ISBN 978-3-540-38127-3 (eBook)
DOI 10.1007/978-3-540-38127-3

Library of Congress Catalog Card Number 67-11280

© by Springer-Verlag Berlin Heidelberg 1976

Originally published by Springer-Verlag Berlin Heidelberg New York in 1976
Softcover reprint of the hardcover 1st edition 1976

Contents

Manuscripts will be accepted by the editors:

Professor Dr. *Jack D. Dunitz*	Laboratorium für Organische Chemie der Eidgenössischen Hochschule CH-8006 Zürich, Universitätsstraße 6/8
Professor Dr. *Peter Hemmerich*	Universität Konstanz, Fachbereich Biologie D-7750 Konstanz, Postfach 733
Professor *Richard H. Holm*	Stanford University, Dept. of Chemistry Stanford, California 94305/U.S.A.
Professor *James A. Ibers*	Department of Chemistry, Northwestern University Evanston, Illinois 60201/U.S.A.
Professor Dr. *C. Klixbüll Jørgensen*	51, Route de Frontenex, CH-1207 Genève
Professor *Joe B. Neilands*	University of California, Biochemistry Department Berkeley, California 94720/U.S.A.
Professor Dr. *Dirk Reinen*	Fachbereich Chemie der Universität Marburg D-3550 Marburg, Gutenbergstraße 18
Professor *Robert Joseph P. Williams*	Wadham College, Inorganic Chemistry Laboratory Oxford OX1 3QR/Great Britain

---

SPRINGER-VERLAG

D-6900 Heidelberg 1
P. O. Box 105280
Telephone (06221) 487·1
Telex 04-61723

D-1000 Berlin 33
Heidelberger Platz 3
Telephone (030) 822001
Telex 01-83319

SPRINGER-VERLAG
NEW YORK INC.

175, Fifth Avenue
New York, N. Y. 10010
Telephone 673-2660

Spectra and Bonding in Metal Carbonyls
Part B: Spectra and Their Interpretation*

P. S. Braterman

Department of Chemistry, University of Glasgow, G 12 8QQ, Scotland

Table of Contents

---

* For 'Part A: Bonding', see Ref. (1).

1

I. General

Since chemistry is largely concerned with trends and differences, compendia of spectra are essential working tools for the chemist. In the case of the vibrational spectra of metal carbonyls, many such compendia exist (2). This article is not intended to add to their number, but rather to examine critically the use of such spectra to give information about bonding.

Vibrational spectroscopy plays a special role in metal carbonyl chemistry, and bands in the 2000 cm^{-1} (CO stretching) region are especially informative. These bands are sharp, strong in absorption, and highly sensitive to chemical environment. Moreover, interactions between (M)CO groups are strong, and so different symmetry species are clearly resolved. Thus vibrational spectra in this region give information about bond type and geometry. Metal carbonyl vibrational spectroscopy is for these reasons a well-established routine tool, and an examination of some of the assumptions commonly made by its users seems timely. Non-vibrational spectroscopy of carbonyls has remained a much more specialised topic, for reasons connected with difficulty of interpretation as well as the availability of instruments (2c, Chapter 8), and will not be further discussed here.

II. The Analysis of Vibrational Data

1. Introduction

Frequencies may be thought of as depending on force constants, and force constants as depending on bonding. Thus, hopefully, determining vibrational frequencies in metal carbonyls can give us information about electron distribution. There are two kinds of difficulty in any such programme. Firstly there are difficulties of practice — do we have enough information and what is the best way to process it? Secondly there are underlying difficulties of interpretation. Are force constants truly a measure of bond order, or are other factors involved? How is the bond between a metal and a carbonyl group in a complex affected by the presence of other ligands? To what extent, then, are we justified in using the CO group as a probe? These two sets of questions are logically quite separate, but bear equally on the design and appraisal of experiments. No set of conclusions can be better than the data on which they are based; but equally, there is less point in pursuing precision if interpretation is limited by conceptual (as opposed to merely practical) uncertainties.

2. Frequencies and Force Constants

Metal carbonyl complexes all possess vibrations in three main regions: (i) CO stretches around 2100—1800 cm^{-1} (terminal CO) or 1900—1700 cm^{-1} (bridging CO) (ii) MC stretches and MCO bends around 650—350 cm^{-1} and (iii) CMC bends below 150 cm^{-1}. Any non-carbonyl ligands will also give rise to vibrational modes. It is unusual for these modes to interact directly with those of the metal carbonyl skeleton, although such mixing does occasionally occur in metal hydrides. The 'other ligand' modes give important information in their own right, and may obscure carbonyl skeletal modes in the medium and low frequency ranges, but will not be considered further here.

To speak of a vibrational mode as due to one kind of distortion is to use the "group frequency approximation". This is a device for qualitative classification, and as such is usually perfectly acceptable, although some mixing of MC and MCO modes of the same symmetry is usual, and in substituted carbonyls CMC and CML bonding modes are constrained to mix by geometry. It is usual, in metal carbonyl chemistry, to go further (at least for the CO stretching modes) and use the *quantitative* "energy factoring" approximation (3). This simply ignores mixing between modes of very different frequency and gross type; thus CO stretches are analysed in isolation to produce a set of CO stretching and CO,CO interaction *parameters*.

Such an approach is remarkably satisfactory (Table 1). It is of course true that stretching a CO bond implies compressing an MC bond, and accordingly there is an MC,CO cross-term of $-1/m(C)$ in the G-matrix. As it happens, this effect is minimised by a corresponding cross-term in the F-matrix. This cross-term is understandable in terms of π-bonding changes. Stretching a CO bond lowers the energy of the CO 2π orbital, increases MC π-bonding, and thus reduces the equilibrium metal-carbon distance. As discussed below, this effect appears in the F-matrix as a positive *interaction constant*.

3

Table 1. Effects of anharmonicity and of energy factoring[a])

Species[b])	Quantity	Anharmonic value	Corrected (harmonic) value[c])
Mo(CO)$_6$[d])	$\nu_1(A_{1g})$	2116.7	2140.2
	$\nu_3(E_g)$	2018.8	2037.3
	$\nu_6(T_{1u})$	1986.1	2026.2
	$F(CO)$	1665	1715
	$F_c(CO, C'\,O')$	16	17
	$F_t(CO, C'\,O')$	20	1
	$k(CO)$	1645	1695
	c	26	27
	t	54	35
Mn(CO)$_5$Br[e])	$\nu_1(A_1)$	2137.9	2158.7
	$\nu_2(A_1)$	2001.3	2030.8
	$\nu_9(B_2)$	2085.4	2083.8
	$\nu_{15}(E)$	2052.2	2078.5
	$F(CO)_a$	1635[f])	1679[f])
	$F(CO)_e$	1745	1788
	$F(CO, C'\,O')_{ae}$	14	20
	$F(CO, C'\,O')^c_{ee}$	3	9
	$F(CO, C'\,O')^t_{ee}$	25	7
	k_1[g,h])	1635	1679
	k_2	1741	1788
	d	31	25
	c	19	13
	t	43	25

a) Force constants and parameters in Nm^{-1}, frequencies in cm^{-1}.
b) All data refer to CCl$_4$ as solvent unless stated otherwise.
c) For critical discussion of correction procedure, see text.
d) Ref. (9); CCl$_4$ solvent.
e) Ref. (5); CH$_2$Cl$_2$ solvent.
f) Derived from corresponding harmonic constants (5a) using corrections of (5b).
g) Anharmonic parameters from (27); harmonic parameters evaluated for these using the corrections of (5b).
h) These data refer to hydrocarbon solution, in which frequencies are (27) 2134.0 (ν_1), 2000.8 (ν_2), 2079.0 (ν_9), 2050.0 (ν_{15}). The effects of solvent shift are clearly far smaller than those of the approximations under discussion.

(One would at the same time expect CO 5σ to become less antibonding between C and O, and thus less available for MC σ-bonding; but any such effect seems to be relatively unimportant). There is of course no guarantee that the relevant inter-action constant, or the metal-carbon force constant, will be the same in all metal carbonyls. However, both of these can be expected to vary monotonically with changes in bond type. A high degree of metal-carbon π-bonding will tend to increase the MC force constant, but at the same time it will tend to increase the MC,CO interaction constant. Thus the distortions due to these effects seem likely to vary uniformly, and in ways that tend to cancel each other, along series of related compounds. We safely use trends in energy-factored CO parameters in place of trends in the far less accessible CO force constants of a full generalised valence force field, especially if we think of these parameters as describing complete MCO units, rather than CO groups in isolation (4).

The generalised valence force field contains both force constants and interaction constants. A force constant F_{ii} gives the stress in an internal coordinate i that would be produced by a unit distortion of that coordinate. An interaction constant F_{ij} gives the stress in an unchanged internal coordinate i caused by unit distortion of coordinate j; it is easy to show that F_{ij} and F_{ji} are equal. If lengthening a bond i sets in train processes that lead to a reduction in the equilibrium length of bond j, then F_{ij} is positive. F_{ij} may be thought of as mixing the vibrations of bonds i and j, in much the same way as the resonance integrals of Hückel theory mix wavefunctions (indeed, historically, the role of Coulomb and exchange integrals is modelled on that of force constants and interaction constants).

The energy factored force field for carbonyls contains interaction parameters as well as stretching parameters. Such parameters are invariably required to be positive, for CO groups attached to the same metal, by the experimental finding that symmetric combinations of the individual CO vibrations occur at higher frequencies than similar antisymmetric combinations.

The absolute experimental accuracy with which interaction parameters may be determined is much the same as that possible for stretching parameters. This is because stretching parameters depend on an averaging of the squares of the observed frequencies, while the interaction parameters depend on the differences of these same squares. However, the error in the case of interaction constants is more serious. The range spanned by interaction parameters is smaller; stretching parameters range from 1400 Nm^{-1} to 1800 Nm^{-1} (18 mD/A), while interaction parameters are generally in the range 0—80 Nm^{-1}. An equal absolute error within a smaller total range implies lower resolution.

Metal carbonyls may be divided into two classes. In carbonyls of the first class, the CO stretching modes do not span any representation of the molecular point group more than once, and the forms of the individual modes are, in the energy factored approximation, determined by symmetry only. The carbonyls of the second class are those for which at least one representation is repeated. All carbonyls contained more than one different kind of CO fall into this class, since each different set of CO groups may be combined to give an internal symmetry coordinate belonging to the fully symmetric representation. In carbonyls of the second class, the form of the normal modes of the repeated symmetry depends on the interaction parameters linking the symmetry coordinates concerned, as well as on the difference in the effective stretching parameters for motion along these coordinates. Thus small errors in the stretching and interaction parameters, in carbonyls of the second class, will each affect the calculated form of the normal modes, and any other properties (such as individual bond dipole derivatives; see Section II.9) that depend on these.

The interpretation of interaction parameters is far from straightforward. Calculations on model systems (2c), and comparisons between parameters and force constants in real systems (4), show that the parameters contain several different terms. Thus in octahedral carbonyls, the *cis* interaction parameter, which is smaller than the *trans* parameter, originates mainly from the true *cis* interaction force constant. The *trans* interaction parameter, on the other hand, contains only a small direct contribution, while the main terms are indirect and involve MC,MC' and MC,C'O' interaction constants. Calculations on substituted

5

species [specifically on $Mn(CO)_5Br(5)$] lead to similar findings. It seems likely that *cis* and *trans* interaction parameters are quite different in nature as well as in detailed origin. Arguments based on competition between CO groups for metal π-electrons (6—8) would imply that *trans* interaction constants should be greater than the corresponding constants for the *cis* interaction, simply because mutually *trans* MCO groupings share two *d*-orbitals rather than one[1]).

This is indeed so for the MC,MC' and MC, C'O' constants (9), but, surprisingly, the *trans* CO, C'O' constants appear to be very small (see Table 1)[2]). Moreover, CO, C'O' interaction parameters are sometimes considerable even when the CO groups are on different metal atoms, and in some such cases their value seems to depend on geometrical rather than bonding considerations. In solid arenechromium tricarbonyls interaction parameters between CO groups on different molecules of the unit cell can be as large as those between groups attached directly to the same atom (10). The parameter linking the near-*cis* (45°) equatorial CO groups on different manganese atoms is far greater than the corresponding parameter for the near-*trans* groups (11). Even in the absence of a metal-metal bond, interaction between CO groups on different metals in such bridged species as $[Mn(CO)_4]_2Br_2$ (12) and tetrahedral $[Mn(CO)_3]_4[SR]_4$ (13) is appreciable, but in the latter case at least depends critically on mutual orientation.

All these facts can be explained if the CO, C'O' interaction constant arises directly from the interaction of local oscillating dipoles (14, 15). Orbital following, such as that involved above to explain the MC,CO interaction constant, is expected to lead to a large oscillating dipole, and the infrared intensities of $\nu(CO)$ bands (see below) may be taken as experimental confirmation of this. Like dipoles repel while opposed dipoles attract; this alone should increase the frequency of symmetric combinations of CO stretching motions, while lowering the frequency of the antisymmetric combinations, and give rise to the observed positive interaction constants. Such constants should, as observed, be greater for mutually *cis* groups. As for groups on different atoms, it is difficult to see how important indirect terms could arise; thus the behaviour of the interaction constants may be taken to follow that of the interaction parameters. The various geometric factors mentioned are then explained, while for solids the interaction is as predicted by exciton theory (16)[3]).

3. Filation Curves

The issues raised in any force field analysis can to some extent be evaded by the method of filation curves (17). This method is applicable when data have been obtained for series of complexes $M(CO)_n, M(CO)_{n-1}L \ldots M(CO)_{n-x}L_x$. For each

---

[1]) Considerations of σ-bonding would point to the same conclusion, since mutually *trans* groups share a common *p*-orbital of the metal, while mutually *cis* groups do not.

[2]) However, the very low values currently reported (9) depend in turn on the procedure used (44) for estimating anharmonicity corrections.

[3]) The finding (12), that metal-metal bonding in species $[M(CO)_4]_2[X]_2$ leads to greater interaction between CO groups in different metals, can also be explained by the local oscillating dipole theory. Quite simply, a metal-metal bond draws the CO groups concerned closer together in space.

ligand L, corresponding frequencies are displayed as a function of the degree of substitution. The effect of substitution in raising or lowering a frequency of a given type is then immediately apparent, as is the relative effectiveness of different substituents L. The method is not confined to CO stretching modes, and was in fact used to locate the elusive MCO bonding mode of E symmetry in Ni(CO)$_4$ (18). Ambiguity of filation could be expected to occur quite commonly in the middle frequency range, which frequently contains MC and MCO modes of the same symmetry, but the success of the method is such as to suggest that the interaction of these two kinds of motion is quite small. The method is, unfortunately, least satisfactory for carbonyls of the second class, since for the repeated symmetry types the filation is inherently ambiguous (see Table 2).

Table 2. Filation sequences for CO stretching modes

Species	Point group	Symmetry type				
M(CO)$_4$	T_d	A_1		T_2		
M(CO)$_3$L	C_{3v}	A_1		E		
M(CO)$_2$L$_2$	C_{2v}	A_1		B_2		
M(CO)L$_3$	C_{3v}	$A_1{}^a$		—		
M(CO)$_6$	O_h	A_{1g}	E_g		T_{1u}	
M(CO)$_5$L	C_{4v}	$A_1{}^b)$	B_2		E	$A_1{}^b)$
trans-M(CO)$_4$L$_2$	D_{4h}	A_{1g}	B_{2g}		E_u	
cis-M(CO)$_4$L$_2$	C_{2v}		$A_1{}^b)$	$A_1{}^b)$	B_1	$B_2{}^c)$
mer-M(CO)$_3$L$_3$	C_{2v}		$A_1{}^b)$	$A_1{}^b)$	B_1	
fac-M(CO)$_3$L$_3$	C_{3v}		$A_1{}^b)$			$E^c)$
cis-M(CO)$_2$L$_4$	C_{2v}		$A_1{}^b)$			$B_2{}^c)$
trans-M(CO)$_2$L$_4$	D_{4h}	A_{1g}			A_{2u}	

a) Correlation with higher frequency component of precursor spectra, based on form of energy factored force fields.
b) These modes all correlate to some extent with each other and with modes of all three symmetry types in M(CO)$_6$.
c) These modes correlate with each other and with modes of symmetry types B_{2g} and T_{1u} in M(CO)$_6$.

4. Incomplete Energy Factored Force Fields

The full generalised valence force field (GVFF) is always seriously underdetermined for any carbonyl complex, and before a solution can even be attempted additional information must be sought, typically by isotopic substitution. However, for carbonyls of the first class, the energy factored force field is fully determined provided all CO stretching modes can be found and assigned. This is a much less formidable task than a few years ago. Grating infrared spectrometers are generally available, and can commonly detect bands that are formally forbidden in the idealised symmetry of the complex under study (19).

There is no lack of well-analysed model systems with which to compare a proposed assignment (1). Laser Raman instruments are commercially available, if expensive. Sample photolysis and absorption can occasionally lead to difficulties in Raman spectroscopy, and some bands of interest may be inactive both in infrared and Raman spectra, but infrared combination spectroscopy has been developed as a crude substitute (20, 23).

There remain the carbonyls of the second class. This includes all complexes containing carbonyls in chemically different environments for which, as shown above, Γ_1 is necessarily repeated, as well as a few polymeric carbonyls, such as [Mn(CO)$_3$SR]$_4$ (13) and [Ir(CO)$_3$]$_4$, in which the representation T_2 of T_d is repeated (21, 22).

Faced with an underdetermined problem, one can impose enough constraints on the solution to render it formally determined; one can decide to settle for less than a complete solution; or one can seek additional information. The only useful source of additional information is isotopic substitution, given the difficulties associated with intensity studies; both isotopic substitution and intensity studies are discussed below.

If one decides to settle for less than a complete solution, one can still obtain a great deal of information from the observed frequencies. In particular, one can delineate the range of possible solutions; and one can determine the average force constant. These points are most easily illustrated by a model system, M(CO)$_a$(CO)$_b$, containing two CO groups of different kinds [more realistic situations have been discussed in detail elsewhere (2c), but the algebra is rather tedious]. The energy-factored force field for such a system determines the vibrational frequencies ν_1, ν_2 through the relationships

$$\begin{vmatrix} k(a) - K & i \\ i & k(b) - K \end{vmatrix} = 0 \tag{1}$$

$$K = K(1) \text{ or } K(2) = 4.0383 \times 10^{-4} \ (\nu_1^2 \text{ or } \nu_2^2) \tag{2}$$

Here $k(a)$, $k(b)$ are the stretching parameters for (CO)$_a$, (CO)$_b$, and i is the interaction parameter. The parameters are expressed in Nm^{-1}, and the frequencies in cm^{-1}. Equation (2) is a rearrangement of the harmonic oscillator equation

$$\nu = \frac{1}{2\pi} \sqrt{\frac{k}{m}} \tag{3}$$

with units adjusted and m taken as the effective mass of a CO group.

It may readily be shown that

$$k(a) + k(b) = K(1) + K(2) \tag{4}$$

so that the average stretching parameter k is determined by

$$k = [K(1) + K(2)]/2 \tag{5}$$

In fact, it is always possible to fix the averaged CO stretching parameter, provided only all the frequencies are known and assigned the correct degeneracy (it is not even necessary to know the exact assignments). It is not possible to fix any of $k(a)$, $k(b)$, i separately, but it is possible to display the implications of any choice by the independent parameter method, applied separately by *Manning, Lewis et al.* (23) and by *Bor* (24). In this case, we may choose $k(a)$ as an independent parameter. It is then easy to show that a plot of i against $k(a)$ is a circle, of radius $[K(1)—K(2)]/2$, and centre $k(a) = [K(1) + K(2)]/2$; $i = 0$. The plot of $k(b)$ against $k(a)$ is of course a straight line [Eq. (4)]. The solutions with i negative may be rejected, at least if the two CO groups are attached to the same metal atom. It may be possible to predict that $k(a)$ is least as large as $k(b)$, and this requirement would halve the range of possible real solutions. If quantitative limits can be imposed, by analogy, on $k(a)—k(b)$, then limits on $k(a)$, $k(b)$ and i follow immediately. The independent parameter method may be applied to any of the situations in which the force field has one degree of freedom, arising out of the two-fold repetition of one irreducible representation, usually Γ_1, in the energy factored force field, and in some of these, chemically reasonable assumptions about the relative size of parameters lead to a very useful degree of limitation on the range of possible values. In particular, for systems $M(CO)_5L$ (Fig. 1) and *cis*-$M(CO)_4L_2$ (Fig. 2) we may assume on general chemical grounds that

$$k_2 > k_1 \tag{6}$$

$$t > d > c > 0 \ \text{(for } M(CO)_5L) \tag{7a}$$

$$\text{and } t > c > d > 0 \ \text{(for } M(CO)_4L_2) \tag{7b}$$

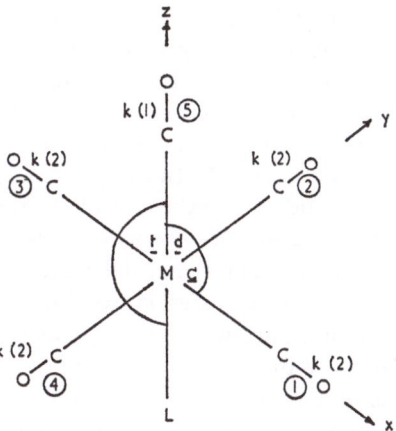

Fig. 1. Axis system and parameters for $M(CO)_5L$

5. Cotton-Kraihanzel and Related Force Fields

It would clearly be advantageous to be able to solve the underdetermined problems of systems $M(CO)_5L$ and *cis*-$M(CO)_4L_2$. Large numbers of complexes of this type have been prepared, and it would be interesting to compare these in as fine a

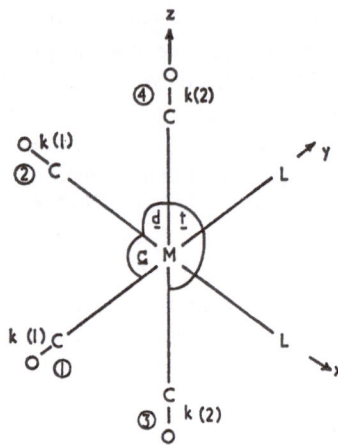

Fig. 2. Axis system and parameters for *cis*-M(CO)$_4$L$_2$

degree of detail as possible. Moreover, the directionality of substituent influences within a single compound can only be examined within these complexes or others of the second class. Frequencies for such carbonyls (or at any rate the frequencies of the A_1 modes) depend on parameters for both types of CO group, and any method of disentangling these influences seems better than none. However, any method must be more or less arbitrary, and the possibility must always be kept in mind that some trends found may be artefacts of the assumptions made.

Two of the methods proposed appear at first sight to be non-arbitrary. In fact, absolute accuracy has been claimed for one of these (25), a refined two-dimensional variational procedure in which all frequencies but one are used as constraints, while this last one is used as a criterion. In fact, it is not possible to fix two unknowns from one parameter, and the method seems to this reviewer an unusually ingenious and elaborate exercise in self-deception, in which the range of possible solutions of the independent parameter method appears to contract to a point.

The second 'absolute' method suggested is the choice of that one solution that makes some chosen interaction parameter as large as possible. The argument is that interaction parameters are increased by π-bonding, that π-bonding increases stability, and that hence, of a range of possible solutions, whichever implies the greatest degree of π-bonding is the most plausible. The argument seems to the present author to be back to front. The force field is a consequence of π-bonding (among other things) rather than a constraint upon it. The argument could imply that the effective stretching parameters for the symmetry coordinates [the diagonal terms in the second order determinants of Eqs. (1) above and (8,9) below] were always equal within any complex. This just happens to be nearly true in Fe(CO)$_5$, one of the successes of this method (26), but would not be expected to hold for species Fe(CO)$_4$L, and is demonstrably far from true for the pentacarbonyl halides and deuterides of manganese and rhenium (27).

The secular equations in the energy factored approximation are, for $M(CO)_5L$,

$$A_1: \quad \begin{vmatrix} k_1 - K & 2d \\ 2d & k_2 + t + 2c - K \end{vmatrix} = 0 \qquad (8)$$

$$B_2: \quad k_2 + t - 2c = K$$

$$E: \quad k_2 - t \quad = K$$

while for cis-$M(CO)_4L_2$ we have

$$A_1: \quad \begin{vmatrix} k_1 + c - K & 2d \\ 2d & k_2 + t - K \end{vmatrix} = 0 \qquad (9)$$

$$B_1: \quad k_2 - t = K$$

$$B_2: \quad k_1 - c = K$$

and for mer-$M(CO)_3L_3$

$$A_1 \quad \begin{vmatrix} k_1 - K & \sqrt{2}\,d \\ \sqrt{2}\,d & k_2 + t - K \end{vmatrix} = 0 \qquad (10)$$

$$B_1 \quad k_2 - t = K$$

Simply to compare frequencies in different compounds is rather similar to setting d equal to zero, while to assume that d is as large as possible is to impose implausible relationships involving k_2-k_1, as well as the interaction parameters. The advantages of imposing some more reasonable set of constraints on the interaction parameters are obvious, and two such sets have been proposed.

The earlier suggestion was made more than ten years ago by *Cotton* and *Krainhenzel* (7, 8), who generalised, and extended to the interaction parameters of substituted species, suggestions earlier made by Jones for the interaction constants of the parent carbonyls (6). The relationships imposed, on the basis of a simple π-interaction valence force field, were, for all three cases of interest,

$$c = d \qquad (11a)$$

$$t = 2d \qquad (11b)$$

The quadratic nature of Eqs. (8–10) ensures that two sets of solutions can be derived; one of these is preferred using the criteria

$$k_2 > k_1 \qquad (12)$$
$$c > 0, \ t > 0$$

These criteria are justified by considering competition between CO and L for d-electrons, and by the orbital sharing theory of interaction parameters, and are shared by the other approximate force field considered.

These assumptions give a degree of overproof, in the sense that the force field can be solved for $M(CO)_5L$ or $cis\text{-}M(CO)_4L_2$ using all but one of the frequencies, and the missing frequency then calculated to within 5 cm^{-1} or so. This is a less impressive feat than it seems, since the force field found would have to be very bad indeed not to give this degree of agreement. It might well have seemed, when Eqs. (11) were first proposed, that one degree of overproof was necessary for species $M(CO)_5L$, in case the B_2 mode escaped detection; for reasons given above this no longer seems a major consideration. Finally, it was claimed that the proposed relationships were an aid to assignment, in the case of $cis\text{-}M(CO)_4L_2$ and $mer\text{-}M(CO)_3L_3$. In species $cis\text{-}M(CO)_4L_2$ it is fairly obvious that the highest frequency belongs to A_1, and the lowest frequency to B_2, but the ordering of B_1 and the lower frequency A_1 mode is not obvious. The actual procedure adopted (8) was to calculate parameters from the highest and lowest frequencies and the B_1 frequency, and to choose the assignment that gave real acceptable roots for the other A_1 frequency. Unfortunately, in some cases, had the chosen assignment been tested using B_2 and the two A_1 frequencies, taking prediction of acceptable values for B_1 as the test, unreal roots would have been found. In other words, the criterion of real roots can be so applied as to be incompatible with the true, as well as the false, possible solution. It has been shown that the difficulty is bound to arise, for many species of this kind, for any treatment incorporating Eq. (17 a), since this constraint requires the selection of a point in an 'independent parameter' plot, outside the range of real solutions. Indeed, it is possible to eliminate t from Eqs. (9), and from inspection of the resultant quadratic form, to derive a minimum value for the ratio d/c if the resultant quadratic is to have real roots. This minimum value is greater than 1 in all cases examined, and can be as high as 1.6 (28).

There remains the claim that Eqs. (11) are an aid to assignment. Suffice it to say that the only ambiguities of assignment [A_1 vs B_1 in $cis\text{-}M(CO)_4L_2$ and in $mer\text{-}M(CO)_3L_3$] can be resolved on simple intensity arguments (B_1 should give the strongest band in the infrared spectrum in both cases), and on the criterion of solvent sensitivity (B_1 should be less solvent sensitive than the lower A_1 mode, since the B_1 oscillating dipole is perpendicular, and the A_1 dipole is parallel, to the molecular dipole and hence to any dipole induced by alignment of a polar solvent). The question can be settled beyond doubt by Raman spectroscopy; both A_1 and B_1 are predicted to be Raman active, but scattering from A_1 should be strong and partly polarised while B_1 should be weak and must be fully depolarised. Where the criteria of infrared intensity and of Eq. (11) are in conflict, it has been conclusively shown by Raman spectroscopy (29) that the former is correct. Thus Eq. (11) as an aid to assignment is either redundant or misleading.

This is not to deny the important historical role of Refs. (7, 8), but rather to advocate the removal of the offending and unecessary constraint of Eq. (11a). As first pointed out by *Paul et al.* (30), it then becomes possible to recognise the existence of more than one kind of π-bonded interaction. The most successful use of arguments of this kind is due to van der Kelen and his collaborators (31, 32). In Refs. (30—32), a parameter y is intorduced to represent the relative π-bonding capacity of CO and the substituent L. This need not be taken too seriously, and the argument can be presented without its help.

We follow Refs. (*31, 32*) in assuming octahedral geometry, and explore the consequences of the (no doubt totally unrealistic) π-interaction force field. For species $M(CO)_5L$, in the axis system of Fig. 1, $CO(1)$ and $CO(2)$ generate an interaction c by sharing $d(xy)$, while $CO(1)$ and $CO(5)$ generate an interaction d by sharing $d(xz)$. $CO(1)$ and $CO(3)$, which are connected by t, share both these orbitals; hence

$$c + d = t \qquad (12)$$

In *cis*-$M(CO)_4L_2$, on the other hand, $CO(3)$ and $CO(4)$ share $d(xz)$ [which links $CO(3)$ with $CO(2)$] and $d(yz)$ [which links $CO(3)$ with $CO(1)$]. Arguing as before, we then have

$$2d = t \qquad (13)$$

and the problems become determinate within the range of real solutions for all cases investigated (*31, 32*)[4].

No extension of the treatment to species *mer*-$M(CO)_3L_3$ has as yet been published. In species $M(CO)_5L$ and *cis*-$M(CO)_4L_2$, it was possible to generate two separate equations connecting c. d, t and y, since there are two kinds of orbitals to be considered. It then becomes possible to eliminate y and generate Eqs. (12, 13). No such procedure is possible for *mer*-$M(CO)_3L_3$, since the mutually *cis* CO groups are linked by only one kind of d-orbital (in the $L(CO)_3$ plane) while the mutually *trans* groups are linked by one orbital in the $L(CO)_3$ plane and by another in the $L_2(CO)_2$ plane. Arguments based on competition for electron density suggest that the latter interaction should be stronger; thus we predict, for this case,

$$t > 2d \qquad (14)$$

but cannot further determine the problem.

Despite this omission, the reviewer feels that the adoption of a constraint is a useful technique for attempting to disentangle the effects of k_1, k_2, c, t and d on the A_1 frequencies, and that the parameters so derived may more usefully be compared than the frequencies from which they are calculated, and are probably good enough for all purposes *except* the determination of the degree of mixing of the A_1 symmetry coordinates and of other properties (bond dipole derivative, bond dipole direction) computed from this. The results derived from the use of constraints are compared with those in which the problem is made determinate by isotopic labelling, in the Appendix. The limited data available bear out the proposed assessment.

6. Isotopic Substitution

Both ^{13}CO and $C^{18}O$ are commercially available, and isotopic substitution has been used in metal carbonyl spectroscopy in two totally different ways. In studies that attempt to correct for the effects of anharmonicity (see II.7. below), sub-

---

4) We must nonetheless take issue with (*32*) for treating $[Mn(CO)_4]_2[Br]_2$ and related species as if they contained isolated $M(CO)_4$ groups, contrary to other evidence (*12*).

stitution must be complete; this is because partial substitution leads to amplitude distributions and anharmonicity corrections quite different from those in the unsubstituted and fully substituted species, which resemble each other. Such a degree of refinement is most often found in attempts to determine the full force fields of simple systems, such as the Group VI hexacarbonyls (9) or nickel tetra-carbonyl (33, 34). In this way, the amount of available information can be increased three-fold, reducing the number of assumptions necessary for a solution of the force field. Moreover, the data for isotopically substituted molecules provide a check on assigned degeneracies, though the *Teller-Redlich product rule* (35). This rule connects the frequencies ω, ω^* of 'normal' and 'labelled' species with the masses m, m^* of the n atoms substituted by labelling:

$$(\omega^*/\omega) = (m/m^*)^{n/2} \tag{15}$$

The product of Eq. (15) runs over all frequencies (degenerate frequencies being repeated) including the notional frequencies for rotation and translation, which can be factored out by standard techniques (3). Equation (15) is a constraint on any set of frequencies to be matched to a force field, and if for any reason the experimental data do not satisfy this equation, attempts to solve the force field may fail to converge. This embarrassing situation is in fact met with in the case of Ni(CO)$_4$ (33, 34). The anomaly is too large to be assigned to simple experimental error or to anharmonicity, and since Eq. (15) is *analytically* true for harmonic frequencies in a field unaltered by labelling, some more subtle effect must be invoked. It has in fact been suggested (34) that the error arises from the equating of absorption maxima with vibrational frequencies. An observed absorption band is in reality an envelope of the separate absorption spectra of groups of molecules with varying amounts and distributions of vibrational energy in the lower energy modes; in other words, an envelope of hot bands. There is no guarantee that the observed maxima for the different bands of a molecular species refer to the same population, still less that the populations for differently labelled species would be otherwise identical. If this argument is accepted we have a fundamental constraint on any attempt to measure vibrational frequencies at all accurately at normal temperatures (for some constraints on low temperature measurement, see Section II.8 below).

For the determination of energy factored parameters, full isotopic substitution is bound to be useless. This is because, in the energy factored approximation, such substitution is predicted to multiply each frequency by a factor of $(\mu/\mu^*)^{\frac{1}{2}}$, where μ is the reduced mass $m_c m_0/[m_c + m_0]$ of carbon monoxide, independent of the values of the parameters concerned. However, a great deal of information can be obtained from the spectra of partially substituted species, for which the Teller-Redlich product rule takes the particularly simple form

$$(\omega^*/\omega) = (\mu/\mu^*)^{n/2} \tag{16}$$

This equation can with advantage be applied separately to each of the vibrational symmetry types of the partially labelled carbonyl, using correlations with the (commonly higher) symmetry of the parent. For example, in a labelled species *trans* $L_2M(CO)(CO^*)_3$, belonging to C_{2v}, derived from a parent of symmetry D_{4h}, we have $\nu^*(B_2)/\nu(E_u) = (\mu/\mu^*)^{\ddagger}$, $\nu^*_1(A_1)\nu^*_2(A_1)\nu^*_3(A_1) = (\mu/\mu^*)\nu(A_{1g})\nu(B_{2g})\nu(E_u)$; these conditions serve as a constraint on any proposed assignment.

A complex containing n CO groups will contain around $n \times 1.1\%$ of species singly labelled in ^{13}CO, due to the natural abundance of ^{13}C. This usually gives rise to a single resolvable 'satellite band' displaced to lower frequency from the strongest $\nu(CO)$ band in the infrared spectrum, usually by around 20 to 35 cm^{-1}. The variation in the degree of displacement may be understood qualitatively by comparing substitution and coupling as perturbations of linked CO groups. If coupling could be ignored, then ^{12}CO and ^{13}CO groups in the labelled species would be vibrating separately, and the labelled species would be displaced to lower frequency by the full amount expected for an isolated CO, *i.e.* by about 45 cm^{-1}. If, on the other hand, the effect of coupling were very much greater than required to produce a displacement of this size, then labelling would scarcely affect the form of the vibrations, which would involve labelled and unlabelled CO groups equally, and the displacement would be equally shared by at least two frequencies; the higher 'satellite' band, however, would be weak in the infrared since it would correspoond quite closely to the symmetric mode of the parent.

Measurement of the position of a single satellite band would in principle suffice to remove one degree of freedom from the energy factored force field. In reality, the experimental uncertainty attaching to such a single band renders this procedure unsatisfactory. The quantity of interest would be the frequency difference between the satellite band and its parent, and this is constrained to lie within quite narrow limits for *any* plausible field. More satisfactory by far is the inclusion of a larger number of frequencies for labelled species, together with the frequencies of the parents, as data for a 'best fit' computer program. This method has the additional advantage that it can be applied even when exact values of some of the parent frequencies are unknown. Indeed, calculation from infrared data for labelled species can be as accurate as Raman spectroscopy in locating infrared-inactive bands (26), is in principle more general, and removes ambiguities of assignment that may exist in Raman solution spectra (11).

Since the natural abundance of ^{13}CO is rather low, it will rarely prove possible to observe all the predicted bands of a singly labelled carbonyl without enrichment. Moreover, such enrichment will remove any doubts as to whether weak bands are due to isotopic labelling or to other causes [such as intrinsic weakness in the spectrum of the unlabelled parent; this is no surprise in 'forbidden' bands activated by solvent packing (36) or by deviations from idealised geometry (19), but can cause confusion in symmetry-allowed bands if these are accidentally weak (37, and references therein)]. The degree of overproof that becomes possible, especially with multiple substitution, helps remove ambiguities implicit in the quadratic or higher equations that arise whenever a representation is repeated, and to guard against errors, due to simple mis-assignment of less full data, that have on occasion led to the publication of wholly erroneous force fields.

Enrichment may be accomplished either thermally or photochemically, depending on the carbonyl involved, or labelled species may be prepared by the reaction of labelled CO with an incompletely carbonylated precursor. Substitution will lead to generation of multiply labelled species by successive replacement reactions. This process is usually governed by simple statistics, although chemically distinct CO groups can be replaced at different rates. Preparation of labelled species from incompletely carbonylated precursors can be less straightforward than predicted; for example, reaction of $Fe_2(CO)_9$ with labelled CO gives a few percent of doubly labelled species from the outset (*38, 39*). Obviously, the possibility of side-reactions, such as varying replacement of ligands by CO, or of CO by impurities acting as ligands, must be borne in mind, especially during photochemical reactions. It must also be emphasised that accurate spectra are essential for all isotopic substitution studies, since the quantities of interest are small frequency shifts. This is even more true for studies directed at the determination of full force fields, since the attempt to improve on the energy factored field must depend on the deviation of these shifts from those that the simpler field requires.

Implicit in the preceding sentence is a fundamental criticism of Eq. (16), deviations from which have been observed (*40*). The effect is dramatically illustrated by the simple model system, $CoCl_2(CO)(PEt_3)_2$ (*41*). Here Eq. (16) requires a shift of the unique parent $\nu(CO)$ band by 48.2 cm^{-1} on substitution with $C^{18}O$, and by 44.6 cm^{-1} on substitution with ^{13}CO. The observed shifts are 44.2 and 44.7 cm^{-1} respectively; thus the shifts are far closer together than expected, and indeed appear to be in the wrong order. It is true that anharmonicity can reduce isotope shifts (see II.7 below), since the labelled species have slightly lower amplitudes of vibration than the parents, but this effect should apply almost equally to both labelled species. The explanation is certainly to be sought in the coupling between $\nu(CO)$ and other motions. Oxygen moves slightly less than the energy factored force field predicts, while carbon moves slightly more. Fortuitously, the deviation from Eq. (16) is very small for labelling at carbon. It has been detected in polycarbonyls, and a possible convergence problem eliminated by taking μ^*/μ as an empirical parameter. However, the deviation from ideality is not large, and even quite ambitious energy factored force field calculations using the calculated value for μ^*/μ (*42, 66*) show an impressive degree of self-consistency.

7. Anharmonicity

The treatment up to this point has assumed that the observed frequencies of a molecule correspond exactly to the classical frequencies for vanishingly small oscillations round the potential energy minimum. For this *harmonic oscillator approximation* to hold for the quantised energy levels of real molecules, it would have to be the case that the potential energy curve corresponded accurately to a parabola over the amplitude of vibration in the vibrationally excited state. This is just not true. The force constant of a bond falls with internuclear distance, becoming zero at infinite separation. As a result, real energy levels are closer together than the harmonic oscillator model requires, and the effect becomes more serious at higher frequencies.

For a simple diatomic molecule, such as CO itself, the desired 'harmonic' or 'mechanical' frequency ω may be found by fitting the energy levels to an expression

$$E = E_0 + \omega(v + \tfrac{1}{2}) + X(v + \tfrac{1}{2})^2 + \text{higher terms} \qquad (17)$$

In Eq. (17), E_0 is the bond energy at equilibrium, uncorrected for zero point energy, v is the vibrational quantum number, zero in the ground state, and X is negative. The error involved in using the observed fundamental frequency ν in place of the harmonic frequency ω is fairly small but not negligible. For example, in $^{12}C^{16}O$ itself (43) $\nu = 2143$ cm^{-1}, $\omega = 2170$ cm^{-1}. Thus the apparent force constant is 1855 Nm^{-1}, as against a mechanical force constant of 1902 Nm^{-1} [5], an error of $2\frac{1}{2}\%$. As well as affecting the apparent force constant, anharmonicity has a slight effect on the isotope shift. Thus for $^{13}C^{16}O$, $\nu = 2096$ cm^{-1}, whereas Eq. 16 requires it to be 2094 cm^{-1}. However, the value found for $\omega(^{13}C^{16}O)$ corresponds to within 1 cm^{-1} with that calculated from $\omega(^{12}C^{16}O)$.

For metal carbonyl complexes, anharmonicity gives rise to more serious consequences, although the effect appears to be fairly small for all motions other than CO stretching vibrations. It is customary to discuss differences in $\nu(CO)$ between related species of as little as 5 cm^{-1}, and implicit in any such discussion is the belief that any differences in anharmonicity are smaller than this. It can be argued (4) that the observed frequencies, corresponding to some kind of force constant averaged over the amplitude of a molecular vibration, give as good a picture of the bonding as do the mechanical frequencies, that correspond to force constants at equilibrium distance. This is no doubt true for very closely related species, in which the forms of the CO modes are similar. There is no reason to suppose that the argument applies to species of different coordination geometry or degree of substitution, in which the forms of the CO modes are necessarily different.

If the effects of anharmonicity on a given vibration are serious, even more serious are the effects on the separation between different vibrations in the same molecule. Equation (17) must be replaced by the generalised form

$$E = E_0 + \sum_i \left[\omega_i(v_i + \tfrac{1}{2}) + X_{ii}(v_i + \tfrac{1}{2})^2\right] + \sum_i \sum_{j \neq i} X_{ij}(v_i + \tfrac{1}{2})(v_j + \tfrac{1}{2}) +$$

higher terms (18)

where the summation in i and j runs over all modes including the separate components of degenerate modes. The large number of parameters involved necessitates observation and assignment of fundamental modes, and of binary and ternary infrared-active modes. This feat has been accomplished for the Group VI hexacarbonyls, and for Mn(CO)$_5$Br, in carbon tetrachloride solution (5, 44). The results are summarised in Table 1, and it is evident that the effects of anharmonicity both on the absolute frequencies of the fundamentals and on the frequency differences between them are very serious.

Unfortunately, while the data of Table 1 demonstrate the seriousness of the problem, they do not go very far to solve it. Firstly, there is the trivial observational point that extension of the treatment to other systems is liable to be extremely difficult. Carbonyl ternaries are weak, and could be masked by C—H binaries; thus molecules with ligands containing CH bonds cannot be examined at all, and the study of other molecules is restricted to their solutions in carbon

---

[5] 18.55 mD/A as against 19.02 mD/A.

disulphide and carbon tetrachloride. These solvents, in particular the latter, react with many carbonyl complexes either thermally or photochemically. Since combination and overtone data are useable only in conjunction with full fundamental data (which must on occasion include Raman spectra) in the same solvent, the problem of photochemical reactivity is not trivial.

Secondly, and most seriously, the validity even of the 'harmonic frequencies' of Table 1 may be questioned (45). The observed binary and ternary bonds are all of symmetry class T_{1u} (in the hexacarbonyls) or A_1 or E (in the case of $Mn(CO)_5Br$), and these symmetry classes are repeated several times both in the fundamental and in the ternary region. Thus we have satisfied the conditions for Fermi resonance. Of course, to show that Fermi resonance is symmetry-allowed is not the same as showing that it occurs, but there is every reason to suspect it in the present case. The physical origin of anharmonicity lies in the existence of direct or crossed cubic and quartic terms in the potential energy expression[6]).

The number of such independent terms in a metal hexacarbonyl is 13 (10 if we discard quartic terms containing the distortion of some CO group raised to an odd power), in addition to the three harmonic force and interaction constants. Thus the number of physical quantities exceeds the number of parameters that may, with the available data, be fitted to Eq. (18). There is the further possibility that the observed frequencies are distorted by interaction with solvent in a way that is not adequately compensated for by Eq. (18). The classical amplitude of a triply excited oscillator is greater than that for one that is only singly excited, and so jostling of solvent and solute molecules, and variability and asymmetry in the solvent sheath, may become important. This may explain the observation that binary and more especially ternary i.r. bands are considerably broader than are fundamentals in the same solvents.

The reviewer is led to a melancholy conclusion. If the theory used to correct for anharmonicity is questionable, and the data are never sufficient to supply overproof, then anharmonicity remains a major source of uncertainty, Indeed, since corrections due to anharmonicity are as large as the errors caused by neglecting the distinction between CO force constants and parameters, there seems little point, as far as the CO vibrations are concerned, in attempting the fuller force field analysis at all.

8. Solvent and Environmental Effects

If we are concerned with the bonding in an isolated molecule, then ideally we should concern ourselves with vapour phase spectra. Unfortunately, only a handful of the species of interest are volatile enough for this to be possible. In any case, vapour phase spectra are complicated by broadening due to unresolved rotational fine structure, and are thus greatly inferior in quality to solution spectra,

---

[6]) Quartic terms cannot be neglected relative to cubic. It is true that they represent a higher order of the potential energy expression. However, first order terms of type $\int \psi_i \cdot P \, \psi_i \, d\tau$, where the ψ_i are eigenfunctions of the harmonic potential energy and P represents deviations from anharmonicity, vanish when P is a cubic (or any odd powered) term but not when P is a quartic (or any even powered) term.

in which the rotational structure is suppressed. For these reasons the vast bulk of metal carbonyl spectroscopy has been carried out in solution; it follows that bands observed are due to energy changes in solute molecules *and their environments*. Solvent and other environmental effects, especially on $\nu(CO)$ bands, are far from trivial (Table 3). One obvious consequence is that frequencies to be compared must refer to the same solvent; a less obvious corollary is that differences in frequencies even in the same solvent may be distorted from the differences in the vapour phase.

We may discuss solvent effects under four headings:

(a) variability and asymmetry,
(b) bulk dielectric effects,
(c) local multipole effects,
(d) specific interactions.

Whatever the nature of the solvent-solute interaction in the ground state, we can expect the packing of solvent molecules round the various molecules of a particular species in solution to show some random variability. In addition, there

Table 3. The effects of environment on frequency

Substance	Solvent	Frequencies (cm^{-1})				
		A_{1g}	E_g		T_{1u}	
Mo(CO)$_6$	(vapour)	2120.7	2024.8		2003.0	(9)
	H[a]	2116.5	2018.9		1989.3	(83)
	CCl$_4$	2116.7	2018.8		1986.1	(9)
	(solid)	2113.6	2005.2			(9)
		A_1	A_1	B_1	B_2	
Mo(CO)$_4$tmen[b]	H	2014	1888	1881	1856	(84)
Mo(CO)$_4$en[b]	CH$_3$NO$_2$	2015	1864	1890	1818	(8)
Os(CO)$_4$I$_2$	H	2164.8	2085.8	2100.8	2050.0	(47)
	CH$_3$NO$_2$	2175.5	2101(sh)	2106.4	2065.5	(47)
Mo(PDMA)(CO)$_4$[c]	H	2026	1923	1938	1914	(47) [d]
	CHCl$_3$	2024	1918	1934	1894	(47) [d]
cis-Mo(PDMA)$_2$-	H		1887, 1828			(47) [d]
(CO)$_2$[c]	CHCl$_3$		1859, 1786			(47) [d]
cis-W(PDEP)$_2$[e]	H[f]		1866, 1807			(85)
	C$_6$H$_5$NO$_2 \cdot 063$[g]), in H		1856; 1803, 1794			(85)
	C$_6$H$_5$NO$_2 \cdot 124$, in H		1852; 1803, 1793, 1774			(85)
	C$_6$H$_5$NO$_2 \cdot 560$, in H		1842; 1773			(85)

a) H = hydrocarbon.
b) en = ethylenediamine; tmen = N,N,N′, N′-tetramethylethylenediamine. Substitution of en by tmen has less effect than the change of solvent which is responsible (29) for the change in order of B_1 and lower A_1 frequencies.
c) PDMA = o-phenylenebis(dimethylarsine), (diars), o—C$_6$H$_4$(AsMe$_2$)$_2$.
d) and data cited therein.
e) PDEP = o-phenylenebis(diethylphosphine), o-C$_6$H$_4$(PEt$_2$)$_2$.
f) In this sequence, H is *n*-hexane.
g) mole fraction.

is no guarantee that the solvent sheath in its lowest possible state, let alone in the states actually met with, will show the same symmetry as the solvent molecule; thus solvent effects will give rise to local asymmetry which may itself in fluid solutions be highly variable.

In low temperature matrices, variability gives rise to a form of matrix splitting, corresponding to the presence of discrete kinds of sites, while asymmetry gives rise to site group splitting, due to lowering of the symmetry of the potential field in one particular kind of site (46). In room temperature solution studies, variability of environment may be expected to affect all bands, leading to a general broadening.

Asymmetry of environment will not affect all bands equally. Common sense suggests, and theoretical analysis ((2) c, Chapter 3) confirms, that the effect of variable asymmetry depends on the form of the mode concerned. Degenerate modes are the most broadened, since asymmetry of environment gives rise to incipient splitting. Non-degenerate modes localised in part of the molecule are less affected, while modes spread over the entire molecule, such as the fully asymmetric carbonyl 'breathing mode' are affected least of all. These arguments apply whatever the particular nature of the solvent-solute interaction, and could explain the common observation that high frequency asymmetric modes, when allowed in the infrared spectrum at all, are distinguished by their sharpness.

The more powerful the solvent-solute interaction, the more pronounced will solvent broadening be; for this reason, saturated hydrocarbons are preferred as solvents for spectroscopy, and such strongly interacting media as methylene chloride and chloroform are to be avoided. It is obvious that the requirements of spectroscopy and those of solubility are in direct conflict. Carbon tetrachloride and carbon disulphide are often used as compromise solvents[7]) (although both of these react thermally or photochemically with many carbonyl complexes) but are generally inferior spectroscopically to alkanes.

The author has found methylcyclohexane a particularly useful solvent, either on its own or (in low temperature glass studies) mixed with isopentane; presumably this is because of poor solvent molecule packing.

Bulk dielectric effects are expected to be important for modes that give rise to high oscillating dipoles. In solvent of high electronic polarisability, giving rise to a large high frequency dielectric constant and refractive index, an oscillating electric dipole in the solute will be stabilised by interaction with induced dipoles in the solvent. Interactions of this kind should lower frequencies in proportion to the intensity[8]) of their associated infrared bands. There will also be multipole-induced multipole effects (e.g. quadrupole-induced quadrupole effects related to Raman intensities) but presumably these will be far smaller. The permanent solvent molecular dipole, which of course contributes to the low-frequency dielectric constant, should have no effect, since orientational relaxation is slow on

---

[7]) Of course, these solvents transmit through much larger regions of the spectrum than do hydrocarbons. Where both $\nu(CO)$ and other modes are to be studied, spectra should if possible be obtained in the $\nu(CO)$ region in alkane (for resolution) and also in the solvent chosen for the other regions (for comparison).

[8]) Both infrared intensity, and the dipole-induced dipole interactions discussed here, depend on the square of the local oscillating dipole.

the relevant timescale, and the time-independent expectation value of an oscillating dipole is zero.

Local multipoles do not, therefore, interact with oscillating dipoles. They do, however, interact with and re-enforce *permanent* dipoles. The sign of the effect so generated on the infrared spectrum will depend on the particular case, and of course different modes will be affected differently. Particularly relevant is a study (*47*) carried out on $Os(CO)_4Br_2$ and $Os(CO)_4I_2$. In this, it was found that CO stretching frequencies *increase*, with increasing solvent multipole effect, in the order

$$C_6H_{12} < C_2Cl_4 < CCl_4 < CHCl_3 < CH_2Cl_2 < CH_3NO_2$$

Moreover, the increase is most marked for those frequencies that are localised in the CO groups *trans* to each other, rather than for these *trans* to halogen; in the notation of Figure 2, k_2 is more affected than k_1.

Any increase of this kind must be due to changes in bonding within the carbonyl complex brought about by the solvent. Presumably the metal-halogen σ-bonds are strongly polarised in the sense $M \rightarrow X$. Solvent dipoles or local multipoles will tend to align so as to reinforce this dipole. The effect will be to increase the effective electronegativity difference between M and X. At first sight, one would expect the $M-X$. σ-bond to be the most perturbed by the solvent field. This would lead to a fall in electron density as the metal and an increase in $\nu(CO)$ frequencies, as observed, but the effect (like the effect of replacing I by Br) should be if anything greatest for the CO groups *trans* to halogen; this is the opposite to what is observed. The same is of course true for the effects of the solvent field on the electronegativities of the different kinds of CO group. The original authors explain the directionality of the observed effect in terms of competition between the various CO groups for π-electron density originating from the halogen (*47*). It is not clear that the currently available rationalisations could have been used to predict the observed effects, but the conclusion has been made more plausible by more recent calculations (*48*) that demonstrate a direct interaction between CO and *cis*-halogen in species such as $Mn(CO)_5X$ and *cis*-$Mn(CO)_4X_2^-$.

So far we have considered solvent effects in terms of gross physical interactions, thinking of the solute as if it were in a cavity surrounded by a more or less homogeneous material. This is of course not true on the molecular level, and the possibility arises of specific interactions between solute and functional groups of the solvent, if the same kind as, though weaker than, what would normally be described as a chemical bond. The coordinated carbonyl group could be expected to be prone to such bonding both at carbon and at oxygen.

Carbonyl oxygen may be capable of taking part in weak hydrogen bonding, especially to chloroform, when the electron density in the coordinated CO is high. The much more dramatic effects (Table 3) of chloroform on the spectrum of *cis*-$Mo(diars)_2(CO)_2$[9]), as compared with $Mo(diars)(CO)_4$, suggests that this is in fact taking place. Such hydrogen bonding would also account for the far greater

---

[9]) For data and references see Table 3. Diars = *o*-phenylenebisdimethylarsine; tmen = N,N'-tetramethylethylenediamine,2,5-diaza-2,5-dimethylhexane; PDEP = *o*-phenylene-bisdiethylphosphine.

influence of chloroform as solvent on those CO groups *trans* to nitrogen in Mo(CO)$_4$ · tmen[9]). The suggestion of specific complex formation receives further support from studies of the spectrum of *cis*-W(PDEP)$_2$(CO)$_{\overline{2}}$ in *n*-hexane-chloroform mixtures; data for this system strongly suggest the formation of two distinct species. The reviewer would suggest that these may correspond to species in which successively one and two CO groups are hydrogen bonded to chloroform. It would be of interest to search for species that showed specific complex formation with chloroform at lower chloroform concentrations than does W(PDEP)$_2$(CO)$_2$, and to test for effects on the chloroform proton magnetic resonance signal.

Carbonyl carbon has been postulated as a possible general site for nucleophilic attack on metal carbonyls (*49*). Indeed, it is difficult to see how else S_N2 attack on the metal hexacarbonyls, a well-attested process (*50*), could take place, and the products of such attack by organolithium reagents are well-characterised intermediates in the conversion of ligand carbonyl to carbene (*51*). It is therefore of interest that pronounced effects have been observed in the low-temperature spectra of Fe(CO)$_5$ in MTHF (2-methyltetrahydrofuran) glass at 77 K, and even in hydrocarbon glasses molar in MTHF (*i.e.*, with a 'mole fraction', of solvent oxygen to solvent carbon, of around 1/50). The effects are concentrated chiefly in those modes derived from the equatorial carbonyl vibrations, and are accompanied (*52*) by changes in the region associated with the in-plane M—CO (equatorial) bending modes.

Environmental effects in solids are much more drastic than those in liquids. The *site group* symmetry of an individual molecule in its environment in a crystalline solid is commonly lower than the symmetry in the gas phase or the effective symmetry in solution. Often more than one molecule is present in the primitive unit cell and the different molecular positions are related by operations of the crystal's space group, such as screw axes or glide planes. In this situation, the spectrum of the solid must be discussed in terms of the *isomorphous point group*, derivable from the *factor group*[10]) of the crystal by simple, well-known procedures (*53*). This 'factor group analysis' gives rise to selection rules, and predicted splittings, that take account of all possible symmetry-allowed interactions between CO groups on molecules in different sites. It may happen that some of these interactions are negligible. In this case, the degree of splitting may be less, and the selection rules more rigorous than the factor group analysis suggests. The material will behave as if under the influence of a group, for which the term *situs group* has been proposed (*54*), of higher symmetry than the actual factor group. Thus attempts to infer or decide between different possible space groups on the basis of vibrational spectra could lead to error. However, considerable progress has been made in the analysis of the vibrational spectra (infrared and Raman) of solid carbonyl complexes (see *e.g. 54—56, 10*). From the polarised Raman spectroscopy of crystals, it is possible to associate particular modes with particular elements of the scattering tensor, to assign frequencies in the isomorphous point group, and to correlate these with frequencies observed in solution;

---

[10]) This is a sub-group of the space group, from which the effects of simple translations have been 'factored off'. 'Isomorphous' means here 'isomorphous to the factor group'.

this technique has been applied with particular success to $Mn_2(CO)_{10}$ (55), the vibrational spectrum of which could not readily be assigned from solution data of the naturally occurring species alone. (The correct assignment had in fact been made (11) from data using the infrared spectra in solution of isotopically labelled species). It has proved possible in simple cases to calculate energy factored force fields for solids, in which interaction parameters may be calculated linking CO groups on molecules in different sites. These are not negligble, and are indeed as large as *intra*molecular interaction parameters.

The effects of parameters linking CO groups on different translationally equivalent molecules (*i.e.* on molecules similarly situated within different unit cells) are not accessible, and are assimilated into the apparent stretching parameters. Thus it appears that, while the spectroscopy of solid carbonyls is of deep interest, the relevance of the data obtained to bonding in the isolated molecule is at best indirect.

9. Intensity Studies

The considerable literature on this topic has recently been reviewed, and the main theoretical models available briefly discussed (57).

It has been known for many years (58) that specific intensities of metal carbonyl infrared ν(CO) bands (total integrated[11]) intensities per CO group) are very high, some 40 times higher than those of organic carbonyls, which are themselves generally regarded as strong (60). This is at first sight a surprising result, since the specific intensity of carbon monoxide itself is low. It is generally agreed [see *e.g.* (48, 49)] that the local dipole within coordinated carbon monoxide, unlike that in organic carbonyl groups, is low. Thus the high intensities observed cannot be due to the motion of atoms carrying high fractional changes, and must be explained by orbital following effects.

It has long been realised that orbital following is an expected consequence of π-back-bonding (61). As a CO group is stretched, the energy of CO (2 π) falls and back-bonding is predicted to increase, giving rise to a large oscillating dipole. Precisely this process is invoked to explain interaction parameters (see Sections II–2, II.4, II.5 above), and it would seem reasonable therefore to hope for a correlation between interaction parameters and specific intensities. Such a correlation seems to exist in the series $V(CO)_6^-$, $Cr(CO)_6$, $Mn(CO)_6^+$, along which frequencies rise and interaction parameters and intensities both fall with increasing charge on the central metal nucleus (the fall in interaction constants between $Cr(CO)_6$ and $Mn(CO)_6^+$ is inferred from the fall between $W(CO)_6$ and $Re(CO)_6^+$, and the general spectroscopic similarities between hexacarbonyls in the first and third transition series) (62, 63). A more extensive test of the proposed correlation has been carried out by examination of two series of related compounds, namely substituted nickel and molybdenum hexacarbonyls (64). As predicted, the higher

---

[11]) Maximum extinction coefficients will not do, even for non-rigorous discussions, since band widths differ so widely. Quantitative discussions should include corrections for instrumental broadening; the procedures available are described in (59).

the interaction parameters, the higher the intensity[12]). Moreover, complexes $Ni(CO)L_3$, for which no interaction parameter can be defined, generally show higher specific intensities than related species $Ni(CO)_{4-n}L_n$ ($n < 3$). Thus the relevant orbital following is that within each individual MCO unit, rather than between these.

As far as bonding is concerned, intensity has generally been regarded as a phenomenon to be explained with help of other data. In structure determination, however, it is a tool in its own right. We can associate with each vibrating CO group a local oscillating dipole, and when only one kind of CO group is present can express resultant relative intensities in terms of the angles between these. When more than one type of CO group is present, it is necessary to know the contribution of each of these to the resultant normal modes; this in turn requires the energy factored force field to have been solved. In favourable cases it is then possible to calculate the relative sizes of the different local oscillating dipoles, using standard methods of linear algebra (3, 57, 65). Moderate errors in the assumed force field, such as those associated with the Cotton-Kraihanzel method, lead to angles considerably different from, and less reasonable than, those found using 'exact' energy factored force fields. Recently, the method has been applied to matrix isolated photolytic fragments, such as $Mo(CO)_5$ (42) and $Fe(CO)_4$ (66), the structures of which could be determined in no other way. These matrix studies embody the results of both low and high degrees of isotopic labelling. The degree of self-consistency and overproof achieved both for the frequencies and for the relative intensities of the bands involved is impressive; so much so as to provide the best possible demonstration of the usefulness of the energy factored force field.

The local oscillating dipole model is analytically valid, and the intensity analysis involved may be regarded as an operational definition of the bond dipole derivatives involved. To equate the direction of an oscillating dipole with that of a CO group is, however, to make an assumption that is less easily verified. This assumption has in fact been questioned several times in the literature. *Kettle* and *Paul* (67) have suggested that, in octahedral derivatives $M(CO)_5L$, there is no reason why the local oscillating dipole of an equatorial CO group should be precisely collinear with the group itself. Unfortunately, any attempt to detect noncollinearity must depend on independent structural data, on the assumption that there is no structural shift in going from the situation in which structure is determined (typically, a crystalline solid) to solution, on the validity of the energy factored force field used, and on the precision of the relative intensity measurements. In none of the cases investigated was it possible to demonstrate noncollinearity unequivocally, and indeed the conclusion seems to be that in the systems studied the effects, if any, are small.

Somewhat earlier, *Cotton* (68) had applied what appeared to be a rather different argument to the infrared intensity distribution in $Mn_2(CO)_{10}$. This species is unique among species $M(CO)_5L$, in that the higher frequency 'A_1' band is more

---

[12]) This conclusion is rigorous for the nickel carbonyls $Ni(CO)_nL_{4-n}$ ($n = 0-2$). Results for the molybdenum series are internally consistent but not directly comparable. They depend on the use of Cotton-Kraihanzel rather than 'exact' parameters, and involve comparison of t (where mutually *trans* CO groups exist) with $2c$ (where they do not).

intense than that at lower frequency. Generally, the reverse is found. Higher frequency A_1 corresponds primarily to an in-plane breathing motion of the equatorial CO groups, the local oscillating dipoles almost cancel, and the band is weak. What intensity there is, derives from the combined effects of non-planarity of the equatorial CO groups (or, more strictly, of their dipole derivatives), and from mixing of the equatorial breathing mode with the stretching motion of axial CO. Indeed, these two contributions are in general opposed; hence the almost vanishingly weak high frequency A_1 band in $HMn(CO)_5$ (37, 28, 65). Cotton's explanation for the anomaly of $Mn_2(CO)_{10}$ is interesting. The motion of interest is not really a localised A_1 mode of one $Mn(CO)_5$ fragment. On the contrary, it is the out-of-phase combination of these modes over the two $Mn(CO)_5$ moieties, and is of symmetry B_2 in D_{4d}. The in-phase combination occurs at considerably higher frequency, and is of symmetry A_1 in D_{4d} and infrared-inactive (55). It follows that the interaction of the two $Mn(CO)_5$ moieties is far from negligible, and the best available quantitative force field analysis (11) confirms this. Cotton's suggestion was that the out-of-phase combination of breathing modes causes an oscillating variation in electronegativity between the two ends of the metal-metal bond. Thus there is orbital following within this bond, increasing the oscillating dipole of the mode concerned, A major objection to this theory is that the intensity anomaly is confined to $Mn_2(CO)_{10}$; it is not shared by $Re_2(CO)_{10}$ (69), nor, even more seriously, by species $[R_3PMn(CO)_4]_2$, in which the axial CO groups are replaced by phosphorus ligands (70). The reviewer has suggested (2c, Chapter 7) that the explanation of the $Mn_2(CO)_{10}$ anomaly is more trivial. The force field of $Mn_2(CO)_{10}$ contains so many parameters that the accurate determination of all of them is a task of extreme difficulty (11). Relatively small shifts in the accepted values would greatly alter the degree of mixing assigned to axial stretching and equatorial breathing in the mode of interest, in such a way as to account for its intensity.

The third objection to the simple method of oscillating dipoles is that of *Darensbourg* and *Brown* (71). These authors argue, quite unobjectionably, that orbital following is the response of the valence electrons to a perturbation, namely the displacement of the atoms involved in the vibration. Such a response can be treated by the standard methods of perturbation theory. The size of the effect depends on the existence of excited states of the same symmetry as the vibration of interest, the energy separation between these excited states and the ground state, and the value of the matrix element through which the ground and excited states are connected by the vibration. Consider, for example, a species *cis*-$L_4M(CO)_2$, in which the angle between the CO groups is 90°. The in-phase CO stretching mode is of symmetry A_1, while the corresponding out-of-phase mode is of symmetry B_2. There is no reason to suppose that the availability or effectiveness of excited states of different symmetries should be in any way related. The localised oscillating dipole model would seem to require that the intensities of the A_1 and B_2 CO stretching modes be equal; but the orbital following model leads us to suspect that this need not be so.

Both the models of *Cotton* (for $Mn_2(CO)_{10}$) and the model of *Darensbourg* and *Brown* (for carbonyls in general) can be accommodated to the local oscillating dipole model provided only the assumption of collinearity between a bond and its dipole is dropped. If distorting an MCO grouping causes an electron flow in some other part of the molecule, then the effective oscillating dipole of that grouping is the sum of two components; in Cotton's model of $Mn_2(CO)_{10}$ one of these is almost perpendicular to the direction of the CO group itself. On the more general orbital following model, if A_1 orbital following is in the case discussed less facile than B_2 orbital following, then a unit stretch of one CO group will produce a smaller electron flow parallel to the C_2 axis than perpendicular to it; thus the angle between the local oscillating dipoles will be greater than 90°, and the intensity distributions will reflect this fact. (Incidentally, there is no justification for assuming that A_1 modes derived from chemically different CO groups will give rise to the same dipole derivative, either overall or in each individual band; the energy differences (decominators) of a perturbation treatment will be exactly the same, but the matrix elements (numerators) will presumably be quite different).

In short, orbital following in other bonds, and orbital following of different degrees of efficiency in different directions, are both operationally equivalent to non-collinearity between the local oscillating dipole and the CO group itself. *A priori*, there is no reason to exclude the possibility of such non-collinearity, but the evidence for its occurrence is not compelling. Hence the use of intensity data in structure determination seems acceptable, provided the contributions of each CO group to each normal mode have been correctly estimated. This caveat is of the greatest importance; a second order error in the predicting of the frequencies of mixed modes corresponds to a first order error in the coefficients concerned.

III. The Implications of Vibrational Data

1. Force Constant, Bond Order, and Bond Type

It is generally assumed that force constant is a measure of valence bond order. If so, this would imply that the force constant of a coordinated CO group would be a direct measure of the degree of metal-ligand back-donation.

The force constant of carbon monoxide can be associated with a bond order of three; that of an organic carbonyl (such as formaldehyde) can be associated with a bond order of two, and that of CO in metal carbonyl complexes can be found by intrapolation. Thus (to the extent that the uncertainties of Sections II.1—II.8 can be ignored) we have a facile probe for d-electron delocalisation within a molecule.

Such a programme was indeed put forward some years ago, at what now seems the high point of optimism this field (72). It immediately ran into difficulties in operation, and is in addition open to objections in principle.

An obvious difficulty of operation concerns the choice of scale. The reference point for bond order 3 is, of course, free carbon monoxide. This leads to no great problems, since the force constant is determined uniquely by the frequency[13]). The other reference molecule presents greater difficulties. Formaldehyde would seem an obvious choice, and the force field of this molecule has been investigated with some care. One calculation using data for H_2CO and D_2CO gives force constants ranging from 1320 to 1100 Nm^{-1}, depending on the exact calculational procedure used, with a value of 1272 ± 56 Nm^{-1} from one, highly plausible procedure (73). This agrees well with a more recent value of 1280 ± 20 Nm^{-1} found by calculations using data for ^{13}C-labelled species (74). Unfortunately, the work also shows a marked fall in CO force constant in going from formaldehyde to acetaldehyde to acetone, despite the fortuitous similarity of frequencies; thus corresponding to the preferred value for formaldehyde we have a value for acetone of only 972 Nm^{-1}, while for acetaldehyde the corresponding value is 1077 ± 54, in acceptable agreement with a more recent value (75) of 1101 ± 9 Nm^{-1} (gas phase values throughout). If the changes from formaldehyde through acetaldehyde to acetone are real, then the CO force constant is highly sensitive to changes that do not seriously affect π-bond order. If they are not, our bench-mark is embarassingly unreliable.

An early difficulty in operation was encountered in the case of the $V(CO)_6^-$ anion (62). Here the best estimated values of the CO stretching constant are 1455 Nm^{-1} (anharmonic) or 1495 Nm^{-1} (harmonic). Taking a force constant of 1105—1135 Nm^{-1} for formaldehyde gave a bond order for CO in $V(CO)_6^-$ of 2.47—2.50. Using the higher value for formaldehyde suggested above will of course lower the apparent bond order ascribed (62) to CO in $V(CO)_6^-$. If we assume a linear relationship between force constant and valence bond order then the higher valence bond order assignable to $V(CO)_6^-$, based on the force constant for acetone

---

[13]) Which frequency? Presumably the observed fundamental frequency, if the comparison is to be with bond orders not corrected for anharmonicity. The effects of phase on free CO are slight.

of 972 cm^{-1}, is 2.56. This result is in conflict with all our expectations. The central metal in V(CO)$_6^-$, formally V(-1), may be regarded as the source of six electrons filling the highest occupied t_{2g} molecular orbital. To reduce the CO valence bond order to 2.5, we would need to assign all these electrons entirely to the CO groups; in other words, we would think of V(CO)$_6^-$ as a complex of vanadium in the $+5$ oxidation state with ligands approximating to CO$^-$. This is scarcely a plausible result. More in keeping with chemical common sense, and with the best available calculations (49), would be equal sharing of the relevant electrons between metal and ligands. This would give a valence bond order of 2.75. Something is seriously wrong with the entire attempt to calculate bond orders from force constants, at least as originally presented[14].

The most optimistic response to this situation is to claim that the force constant — π-bond order relationship is still valid, but that the reference points need to be changed; V(CO)$_6^-$ itself is then a possible reference compound (76). The relationship can then only be quantified by using calculated orbital populations for the reference species, and can only be tested by more extended comparisons between calculated bond order and observed force constant. Precisely this test has been applied to a whole group of substituted and unsubstituted octahedral carbonyls of groups VI and VII, the substituents in every case being halide (77). The data used in fact were not force constants, but Cotton-Krainhanzel parameters; this does not actually matter, since no reference molecules were used at all. Excellent agreement was found with an expression.

$$k = 3581 - 1173[2\pi] - 950\,[5\,\sigma]\ \text{Nm}^{-1} \tag{19}$$

when $[2\pi]$ and $[5\,\sigma]$ are the occupancies of these orbitals in each CO ligand. The difference in coefficient between $[5\,\sigma]$ and $[2\pi]$ is on the border of the significant. Unfortunately, it is not possible to test the relationship directly for the unsubstituted hexacarbonyls, owing to differences in basis set (77 a).

If Eq. (19) is applied to CO itself, a parameter of 1681 Nm^{-1} is predicted, in gross disagreement with observation. It could be suggested either that $5\,\sigma$ is for some reason more CO-antibonding in complexes than in free CO; or that the quite gross changes in bond length between free and coordinated CO give rise to changes of yet a different kind; or most plausibly that CO lacks opportunity for the extra orbital following discussed below.

Application of Eq. (19) to V(CO)$_6^-$ ($k = 1455$) (62) or to Cr(CO)$_6$ ($k = 1645$) (9) implies total ($\sigma + \pi$) occupancies of around 2.1 electrons for CO in the former compound, and 1.9 in the latter. These values seem reasonable enough; in V(CO)$_6^-$ back-donation (populating 2π) slightly outweighs forward donation (depopulating $5\,\sigma$) while in Cr(CO)$_6$ the reverse is true. Be that as it may, if the calculations presented (77) have any value at all, *it is not legitimate to ascribe changes in CO parameters, frequencies or force constants to changes in π-bonding alone.* For example, there is little difference in stretching parameter between the CO groups *trans* to halogen in *cis*-Fe(CO)$_4$I$_2$, and those *cis* to halogen in Mn(CO)$_5$Cl, but this effect

---

[14] A quadratic function passing through the points (0,0), (2.972) and (3,1902) does little better, giving an apparent bond order for V(CO)$_6^-$ of 2.58.

is the result of conflicting influences. The occupancy of 2π is 0.1 electrons greater in the iron complex, while that of 5σ is 0.1 electrons lower. Either of these effects alone would have caused a difference of around 100 Nm^{-1} between the stretching parameters of the groups in question, far outside the plausible limits of reliability of the approximations used; in fact the effects are opposed and the difference found is 3 Nm^{-1} [predicted from Eq. (19), 30 Nm^{-1}]. Differences within molecules are also attributable to a combination of effects. For example, in Mn(CO)$_5$Cl, in the Cotton-Krainhanzel approximation, the parameter k_2 is 118 Nm^{-1} higher than k_1. The predicted difference is 127 Nm^{-1}. A separation of 168 Nm^{-1} arises from differences in π-bonding, while differences in σ-bonding alone would have led to a difference of 41 Nm^{-1} in the opposite direction. Axial CO is accepting more electron density than are the equatorial CO groups, but it is also donating more, and to some extent these effects are opposed.

The implications of the calculations of Ref. (77) are more fully drawn out in Figs. 3—5. There is obvious evidence for a correlation between population of the 2π orbital (i.e. back-bonding) and lowering of CO frequency. A correlation between high CO frequency and high 5σ population (poor forward bonding) is also apparent but less striking. The combined relationship of Eq. (19) is clearly

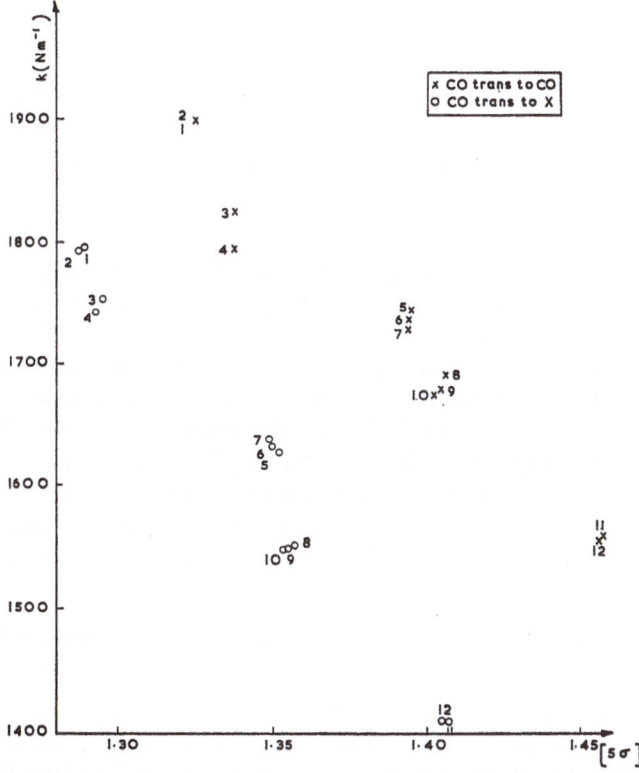

Fig. 3. Cotton-Kraihanzel parameters vs 5σ-orbital occupancy [after Ref. (77)]. (1) Fe(CO)$_4$Cl$^+$; (2) Fe(CO)$_4$Br$^+$; (3) Fe(CO)$_4$Br$_2$; (4) Fe(CO)$_4$I$_2$; (5) Mn(CO)$_5$Cl; (6) Mn(CO)$_5$Br; (7) Mn(CO)$_5$I; (8) Mn(CO)$_4$Br$_2^-$; (9) Mn(CO)$_4$IBr$^-$; (10) Mn(CO)$_4$I$_2^-$; (11) Cr(CO)$_5$Cl$^-$; (12) Cr(CO)$_5$Br$^-$

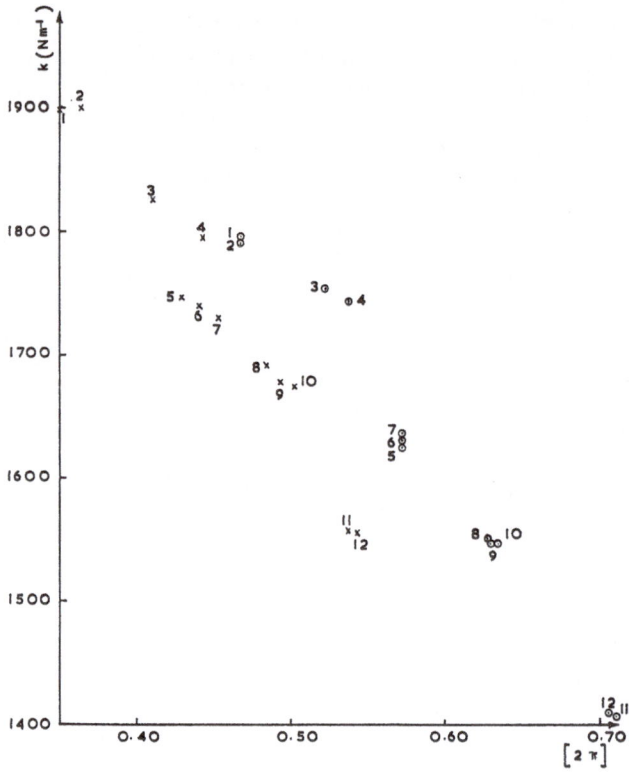

Fig. 4. Cotton-Krainhanzel parameters vs. total 2π-orbital occupancy

superior to a relationship involving either σ-bonding or π-bonding only, and indeed is so powerful as to lend a measure of validation to the approximate force field method itself[15]).

We are now close to being in a position to examine the effects on CO stretching parameter, for the isoelectronic isostructural series of first row carbonyl halides considered, of changes in metal, degree of substitution, coordination site, and halide. Along the sequence Cr(O), Mn(I), Fe(II), both 5σ and 2π populations fall. In other words, a higher oxidation state of the metal renders it at once a better σ-acceptor and a poorer π-donor; perhaps a predictable result. Numerically the π-effect is the greater, but both are in the same direction, and parameters

---

[15] Unless the agreement between m.o. calculations and CO parameters is fortuitous, we must take each of these to be measuring an aspect of reality, although not necessarily the particular aspect claimed. A purely chance agreement can be discounted. A fortuitous agreement could arise if purely artificial computed effects on bonding of change of metal, degree of substitution, and coordination site, all happened in the correct ratio to match the real effects on spectra; but this seems highly unlikely. The force field used ensures fairly well (the van der Kelen force field ensures exactly) that the average of k_1 and k_2 is correct; the fact that data for *cis* and *trans* CO groups lie on the same straight line shows that the difference calculated is also more or less correct.

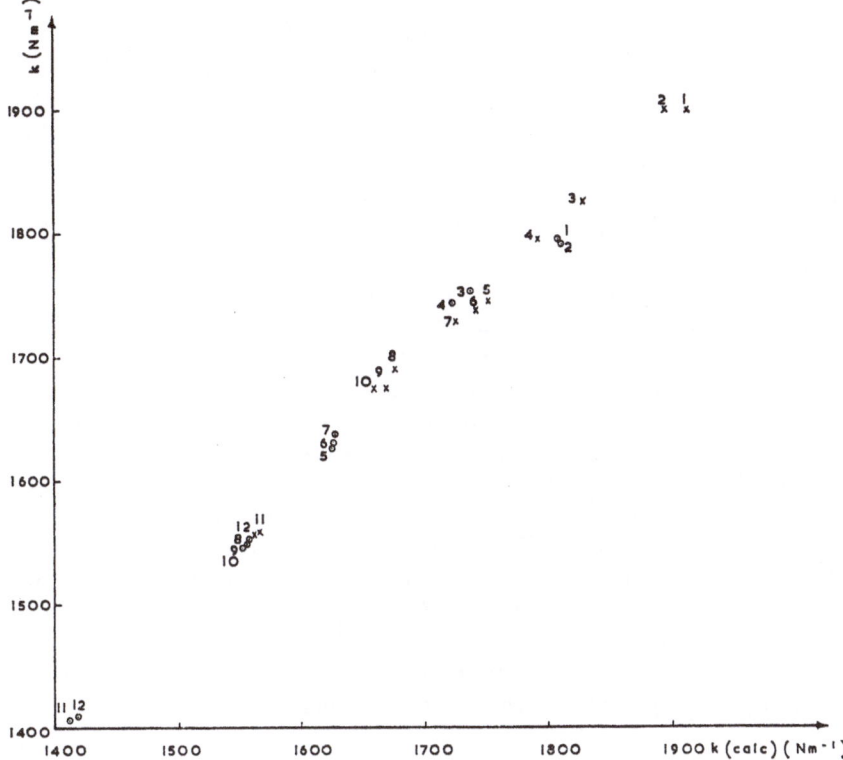

Fig. 5. Cotton-Krainhanzel parameters vs. values calculated from Eq. (18)

increase along the sequence more rapidly than either alone would require. Increased substitution leads to a small increase in 5σ population on the carbonyl ligands, but to a much larger increase in 2π population. This is in accord with the conventional view of CO groups competing with each other for $d(\pi)$-electron density. When we compare carbonyl groups *trans* to CO with those *trans* to halide in the same complex, we find that the latter have higher π-electron densities and also somewhat lower σ-electron densities. The change in π-electron density is precisely as expected, since CO groups compete most effectively when mutually *trans*, and coordinated halide is a π-donor. The change in σ-density seems to show coordinated CO exercising a direction-specific *trans* influence by a σ-mechanism, and by this criterion CO is a stronger σ-donor than halide. This conflicts of course with conventional basicities, but the σ-donor power of CO in complexes is enhanced synergically by π-acceptance. The σ-density changes may also reflect synergic influences more directly; the higher π^*-orbital population of axial CO will tend to make it a better σ-donor on electronegativity grounds. Be that as it may, the differences in σ-bonding reduce the difference in CO parameter between axial and equatorial CO, compared with those that would have resulted from π-electron density changes alone.

31

The effects of changing ligand are slightly less clear, because of the restricted range of ligands considered, but the treatment is remarkably successful in predicting trends even over the short range spanned by change of halide. In most cases [equatorial CO in $Cr(CO)_5X^-$, $Mn(CO)_5X$; all CO in $Mn(CO)_4X_2^-$, $Fe(CO)_4X_2$] CO parameters are required by Eq. (19) to fall with increasing atomic number of X, and in fact do so. In the sequence $Fe(CO)_5Cl^+$, $Fe(CO)_5Br^+$, the calculated effect of changing halogen is very small (1 or 2 Nm^{-1}) and actually opposed in direction to the small difference observed. For the axial CO groups of $Cr(CO)_5X^-$ and $Mn(CO)_5X$, CO parameters are clearly predicted to fall on descending the halogen subgroup, and this prediction is also borne out. In view of these successes, it is worth looking very closely at the origins of the predicted shifts. Sensitivity to change of halogen is greater for CO groups *cis* to halogen than for those *trans*. The bigger single cause of this sensitivity appears to be specific population of that 2π orbital that lies parallel to the metal-halogen bond. In other words, the substituent donates directly to those carbonyl groups *cis* to it. Such donation is greatest for iodide, which accordingly gives rise to the lowest CO parameters. For the axial carbonyl groups of $M(CO)_5X$, the biggest single agent of change is the sensitivity of the carbonyl 5σ population to the nature of X. This falls with increasing atomic number of halogen, while the 2π orbital population remains more or less constant. As a result, the CO parameters increase from chloride to bromide to iodide. This result could not be explained on general grounds of electronegativity, and would appear to indicate a specific σ-*trans* influence which in these compounds is greater for the smaller halogens.

2. The Analysis of Substituent Effects

It is generally agreed that more electron-donating ligands tend to reduce carbonyl stretching parameters. It is also agreed that the overall donor power of a ligand depends on a balance of σ- and π-bond type. There is far less agreement about the relative importance of these two factors, and considerable effort has been devoted to attempts to separate the two types of effect. Apart from the calculations discussed above, the arguments have generally rested on the directionality of substituent effects, with in some cases supporting evidence from M—C stretching frequencies and bond strengths.

In a complex $M(CO)_5X$, k_1 (for axial CO) will depend quite strongly on the degree of π-bonding between M and X. If X is a π-acceptor, then it will in genera be competing with the axial CO for electrons both in $d(xz)$ and in $d(yz)$. Conversely, if it us a π-donor, then it will be repelling these electrons and making them more available. On these grounds, a π-accepting group X would be expected to strengthen the axial CO bond and weaken the axial MC bond. These effects will also be exerted on the equatorial CO groups, but to a smaller extent, and several authors (*78, 79*) have agreed that the effect on the equatorial groups should be half that on the axial group, since each equatorial group shares only one $d\pi$ orbital with Cap. x, while the axial group shares two.

The effects of σ-bonding may be more complex. A good σ-donor would naively be imagined to increase electron availability at the metal, by a simple inductive mechanism, and thus to increase π-donation from metal to CO while reducing

σ-donation from CO to metal. Enough is known about the effect of charge density in isoelectronic series (62) for us to infer that the CO stretching parameters should then fall, while on balance the MC force constants would remain the same, although the frequencies of the individual MC modes would probably change. These effects would presumably apply equally to all CO groups in the complex. At the same time, σ-donation from X could be expected to give rise to two distinct directional effects. Firstly there is the familiar σ-*trans* effect (80), which would be specific to the axial CO group. This would decrease CO to metal σ-donation and thus lower both MC and CO force constants. These effects seem not to have been included in attempts to quantise directional influences in complexes (78, 79). Secondly, there is direct donation from X into the 2π orbitals of the equatorial carbonyls. The significance of this effect in carbonyl halides is apparent from the calculations of *Hall* and *Fenske* (77), and some authors (79) invoke direct donation as the *only* σ-bonding effect. It is difficult to see how this effect would influence the equatorial MC force constant, but it could be that this would be increased by enhanced ligand to metal σ-bonding.

Our knowledge of trends in metal-carbon modes in substituted carbonyls is fragmentary, and there are obvious problems involving mixing of MC stretching and other modes. One study has shown (79) that the equatorial E MC mode is lowered by electronegative groups X, but it is not obvious why this should be so. In the absence of clearer information about metal-carbon modes, we have a typical underdetermined problem. Two simplifying solutions[16] have been proposed. The earlier of these, due to Graham (78), regarded the σ-influence as non-directional. This leads to equations

$$\Delta[k_1] = \sigma + 2\pi \tag{20}$$
$$\Delta[k_2] = \sigma + \pi$$

More recently, *Dobson* (79) has assigned the σ-influence entirely to the direct donor (*cis*) effect. This leads to equations

$$\Delta[k_1] = 2\Pi \tag{21}$$
$$\Delta[k_2] = D^{16)} + \Pi$$

(writing 'D' for the original authors 'd' to avoid confusion with the interaction parameter so denoted).

It would be possible, and in this author's opinion valuable, to carry out model calculations that would help distinguish between the assumptions of Eq. (20) and Eq. (21). For example, fictional species $Cr(CO)_5L$ and $Mn(CO)_5L^+$ could be investigated, in which L is a 2-electron atom with one filled σ-donor and two degenerate empty π-acceptor orbitals, and the effects of altering the size and Coulomb energy of these on orbital populations and energies could be determined. In the

---

[16] The third possible simplification would be to assume the σ-influence to operate predominantly as a *trans* effect, perturbing axial much more than equatorial CO groups. This leads to a totally unsatisfactory order of σ-donor abilities.

absence of such model calculations, we can at least use such calculations as do exist, and also test the assumptions for reasonableness in relation to real compounds.

The most suitable calculations are those of *Hall* and *Fenske* already referred to (77). According to these, the differences in the series $Mn(CO)_5Cl$, $Mn(CO)_5Br$, $Mn(CO)_5I$ are almost entirely due to changes in the population of the 2π (z) orbital of the equatorial CO groups. As shown in Table (4) this is in accord with the model of Eq. (21). which attributes the change in equatorial CO parameter mainly to changes in D, while the values of π vary only slightly. The key experimental finding that k_2 varies much more for these compounds than k_1 is thus readily explained. On the model of Eq. (20), however, a smaller variation in k_1 than k_2 can only be accounted for by a balance of opposed σ and π effects. Cl is calculated to be a poorer σ-donor than I to the same degree that CF_3 is a poorer σ-donor than CH_3, and is constrained to appear a better π-donor than I by a factor of 2. The difference in π-donor ability would operate on the $d(xz,yz)$ orbitals,

Table 4. Graham's σ and π effects [(Eq. (20)] and Dobson's D and Π effects [Eq. (21)] for some species $M(CO)_5X^a$)

Species $Mn(CO)_5X^b$:	$\Delta(1)$	$\Delta(2)$	σ	π	$2D$	2Π
X = SnCl$_3$	80	49	18	31	18	80
SnClMe$_2$	48	25	2	25	2	48
SnMe$_2$Cl	46	— 10	— 66	56	— 66	46
SnMe$_3$	18	— 33	— 84	51	— 84	18
SnPh$_3$	23	— 18	— 59	41	— 59	23
CF$_3$	48	52	56	— 4	56	48
H	31	11	— 9	20	— 9	31
CF$_3$CO	54	41	28	13	28	54
CH$_3$ chosen zero						
Cl	11	68	125	— 57	125	11
Br	15	59	103	— 44	103	15
I	19	46	73	— 27	73	19
AuPPh$_3$	— 42	— 80	—118	38	—118	— 42
W(CO)$_5$L^c):						
L = CO	128	69	10	59	10	128
P(OBu^n)$_3$	68	20	— 28	48	— 28	68
PPh$_3$	39	16	— 7	23	— 7	39
PBu$^n_3$	39	3	— 33	36	— 33	39
NC · CH$_3$	9	18	27	— 9	27	9
Cyclohexylamine chosen zero						
Ar^d)	14	51	88	— 37	88	14
Pyridine^e)	— 30	0	30	— 30	30	— 30
(Pyridine $^{-e}$)	—231	— 95	41	—136	41	—231

[a]) Shifts (Nm^{-1}) in $k(1)$, $k(2)$ from chosen zero. Negative values denote good σ- or π-donating substituent.
[b]) Data from (78); hydrocarbon solvent.
[c]) Data for (79); hydrocarbon solvent.
[d]) Argon matrix [from data of J. J. Turner et al., as quoted in (79)].
[e]) In 2-methoxyethane, from (82).

for which it is almost precisely neutralised by the inductive effect. The observed trend in k_2 should therefore be attributable to an increase in the availability of the $d(xy)$ orbital, reflected in an increase in population of the in-plane 2π (equatorial) orbitals, while that of the out-of-plane orbitals remains steady. This is in complete contrast with the calculations of (77).

The test of physical plausibility is equally harsh to the model of Eq. (20), as the data of Table (4) show. For on such a model, the effect of replacing a pure σ-donor by a ligand vacancy should be distributed equally over k_1 and k_2. In fact, the change from $W(CO)_5NH_2C_6H_{11}$ to $W(CO)_5(Ar)$ affects k_2 very much more than k_1. A number of other differences between the intepretations of Eqs. (20) and (21) also favour the latter. Thus Eq. (21) leads to acetonitrile being classified as a π-acceptor rather than π-donor, and this is in accord with the usual interpretation of $\nu(CN)$ frequency changes. Moreover, Eq. (21) presents a regular decrease in π-acceptor capacity towards $Mn(CO)_5$ along the series Cl_3Sn, Cl_2Me-Sn... Me_3Sn, while Eq. (20) leads to a more irregular ordering with Me_3Sn, paradoxically, a better π-accepting group than Cl_3Sn. The explanation can be advanced that the π-donor effect of chlorine in Cl_3Sn reduces the availability of the $5d$ orbitals of tin more than electronegativity effects increase it; but this is unconvincing in view to the generally accepted fall in π-acceptor power from PCl_3 to PMe_3[17]).

However, it would be too much to hope for either Eq. (20) or Eq. (21) to give a completely satisfactory analysis of trends. Quite apart from the orbital following effects discussed in the next section, either of these equations ignores two significant effects of σ-bonding, as well as assuming that the weighting of π-interactions in k_1 and k_2 is the same on all compounds. Thus even if Eq. (21) is preferable, it does still lead to some embarassing anomalies. Hydrogen is calculated to be a medium strong π-acceptor compared with CH_3, intermediate between Ph_3Sn and $PhCl_2Sn$, and C_6F_5 is closely comparable to H, while the halogens appear as π-acceptors[18]).

It seems that while Eq. (20) is always unsatisfactory, Eq. (21) is fairly satisfactory for the series $W(CO)_5L$ but not for the series $Mn(CO)_5X$. This is hardly surprising, since σ-variation is expected to be more important, and π-variation less important, in the latter series, and the error of treating σ-effects as one-dimensional will accordingly be more serious.

In this section, we have discussed in detail only one class of carbonyls, namely monosubstituted octahedra. More highly substituted octahedra have been less thoroughly investigated, either because the CO parameter analysis is less certain, or because the separation of ligand cis- and trans-influences is less clearcut. In substituted tetrahedra, the distinction between cis and trans CO groups does not exist, while in 5-coordinate complexes, the situation is further complicated, quite apart from the effects of substituents, by the axial-equatorial distinction.

---

[17]) Admittedly, this analogy would apply more powerfully to the series $Cl_{3-x}Me_xSi$; but the data for silicon compounds, while incomplete, strongly indicate that substituent effects at Si and Sn are very similar.

[18]) The only π-donor on the model of Eq. 21 is Ph_3PAu; this is, ironically, a strong π-acceptor on the model of Eq. (20).

3. Orbital Following Effects

In the preceeding section, CO parameters were related exclusively to the electronic distribution at equilibrium. This is not realistic if orbital following is appreciable. Conceptually, one can divide the process of distorting a molecule into two parts. One can imagine a displacement of nuclei while the electron distribution is held constant according to some criterion, such as the ratio of atomic orbital coefficients in an LCAO molecular orbital scheme. One can than allow the electrons to relax to their lowest energy distribution in the field of the displaced nuclei. This process of *orbital following* will lower the energy of the distorted molecule. It follows that the force constants of real molecules, in which orbital following occurs, are lower than they would have been had the form of the electronic wavefunction been held constant (whatever that may mean) throughout the vibration.

There is every reason to believe that such orbital following effects are especially important for metal carbonyls. Such effects have already been invoked (Section II.9) to explain the high specific infrared intensity of CO stretching modes in metal carbonyls, and also at several points to explain the large positive interaction parameters connecting CO groups on the same metal. In other words, CO groups in carbonyl complexes exhibit a degree of *extra* orbital following not to be found in such saturated species as free CO or formaldehyde. Not surprisingly, their force constants are then much lower than any correlation with bond order would require[19]). The effect is massive. Thus the application of Eq. (19) to free CO ($[2\pi] = 0$, $[5\sigma] = 2$) leads to a predicted force constant of 1681 Nm^{-1}, as against 1855 Nm^{-1}. The extra orbital following in the metal carbonyl halides to which Eq. (19) was fitted has lowered them, relative to CO, by 170 Nm^{-1}. This corresponds to more than one third of the entire range of values considered.

The phenomenon would not be too serious if we could presume that the degree of extra orbital following would be the same in all carbonyl complexes, or would at least vary monotonically with bond population. Unfortunately, even this is likely to be untrue. More polarisable ligands, and in particular either π-accepting or π-donating ligands, are likely to form an effective conjugated system with both *cis* and *trans* CO groups; thus equations such as Eq. (19) can only be applied with confidence to very restricted series of compounds. Moereover, the degree of extra orbital following is likely to be totally different in complexes of different geometry (such as tetrahedral *vs.* octahedral complexes), or even in chemically different environments within the same molecule (such as axial and equatorial CO in a trigonal bipyramidal complex). It follows that there is no hope at present of anchoring such disparate groups to the same scale. It may be reasonable to maintain that there is less back-bonding in $Ni(CO)_4$ than in $Cr(CO)_6$ on general chemical grounds or on the basis of calculations; but to claim this on the basis of force constants is to go beyond the evidence.

---

[19]) Professor *Fenske* (77a) has independently reached rather similar conclusions.

IV. Conclusions

We began this review by questioning the conventional use of carbonyl stretching data both on experimental and on conceptual grounds. Our conclusion must be that experimentally, the usual approximate procedures are adequate for most purposes, but that the conceptual interpretations offered are less adequate.

CO stretching parameters behave in much the same way as more rigorous force constants, Interaction parameters are more complicated in origin, but may be used to describe the coupling of MCO units with apparent success. Approximate stretching parameters agree with 'exact' stretching parameters as closely as chemical interpretation requires, although approximate interaction parameters are less useful. Anharmonicity raises unresolved problems for the entire subject and studies directed to that end find the effects to be large and variable from mode to mode. However, partial isotopic labelling studies that completely ignore anharmonicity have been highly successful, although the amplitude distributions of a molecule must be quite drastically altered by such substitution.

Observed bands depend (except for vapour phase spectra) on medium as well as on substrate. This is true even in inert gas matrix studies, and is much more obviously so in solution. This phenomenon may be turned to advantage in the study of solvent-solute interactions, and in any case may often be minimised by careful choice of solvent. Observed intensities confirm simple ideas of orbital following, and intensity distributions may be related to structure in well-understood ways, at least when 'exact' parameters are available, or when only geometric effects are relevant.

There is no prospect of establishing a simple relationship between CO stretching parameter (or force constant) and bond order, still less of breaking down changes in bond order into σ and π components, on the basis of vibrational spectra alone. In all carbonyl complexes, the parameters of interest are grossly lowered by orbital following effects. Whether within the field of metal carbonyls these parallel other effects is not yet known, but certainly they suffice to vitiate comparison between metal carbonyl complexes and less polarisable model compounds. Even ignoring orbital following differences, changes in stretching parameter may reflect changes either in σ- or in π^*-orbital population at CO. The σ-orbital population is sensitive to metal electronegativity and may also perhaps be affected by specific orbital (*e.g. trans* ligand) influences and or by the synergic consequences of π-acceptance. The latter, which tend to predominate, are sensitive to metal electronegativity, to specific orbital effects (in particular, to mutual competition between CO groups, and to competition between CO and other π-accepting ligands), and in octahedral complexes at least to direct *cis*-interactions with substituents. To explain detailed trends due to ligands (for example, trends in $k_2 - k_1$ in substituted octahedral carbonyls) is beyond the power of any single phenomenological model, since the number of variables exceeds the number of observables, and no one model gives consistently sensible results. Thus it does not matter too much if at some later date our CO parameters are shown to be merely mediocre approximations, since even if we had unimpeachable information we could not know how to use it.

The qualitative use of CO frequency or force constant as an electron density probe seems vindicated. More exact interpretation is sometimes possible, provided the bonding is already independently understood. This is of course exactly the opposite to what we would hope for, but attempts to use carbonyl vibrational data as quantitative or specific bond type probes seem doomed to failure.

V. Appendix

The approximate solutions to the energy factored force field offered by the approximations of *Cotton* and *Kraihanzel* (*7, 8*) and by *van der Kelen* and his colleagues (*31, 32*) are compared in Table (5) with the results of isotopic substitution studies. For purposes of comparability, all results are based on the same data; *i.e.* the approximate parameters have been re-calculated for this work from the all —$^{12}C^{16}O$ data of the more exact studies.

The equations were solved analytically after substitution of the assumed relationships of interest into the 'exact' energy factored force field. For the Cotton-Kraihanzel treatment of species $M(CO)_5X$, $\nu(B_2)$ was not used. In the corresponding treatment of species *cis*-$M(CO)_4X_2$, the expressions for $K(B_1)$, $K(B_2)$, and $K(A_1) + K(A_2)$ were used; this procedure means that all the available $^{12}C^{16}O$ data are used, as well as obviating the need to solve a quadratic.

The results are high encouraging. The deviations between Cotton-Kraihanzel and van der Kelen parameters are so small that it does not matter which set is used. Both approximations correctly predict the sequences $CH_3 < Cl < Br < I < D$ for k_1 and $CH_3 < D < I < Br < Cl$ for k_2 in the series $Mn(CO)_5X$, and also, rather surprisingly, the order $I < Br < Cl < CH_3 < D$ for t and c. The mixing parameter $2d/[k_2 + t + 2c - k_1]$ is not correctly ranked, still less correctly predicted, by either approximate method, confirming the criticisms in this review of attempts to derive eigenvectors from 'approximate' parameters. The difference $k_2 - k_1$ is correctly ranked in both approximate treatments for the species $Mn(CO)_5X$, strongly suggesting that the failure of the models used to treat this difference is conceptual rather than experimental. Results for the *cis*-disubstituted compounds are slightly less satisfactory in detail, although the increase in both k_1 and k_2 from $Fe(CO)_4I_2$ to $Fe(CO)_4Br_2$ is correctly shown by both the approximate models.

Acknowledgements. It is a pleasure to thank Professor R. F. Fenske, Dr. J. R. Miller, Dr. K. Noack, Dr. M. Poliakoff and Professor J. J. Turner for helpful communications, and in some cases for access to results before publication.

Table 5. 'Exact', Cotton-Kraihanzel, and van der Kelen parameters[a])

Species	'Exact' Parameters[b]						Cotton-Kraihanzel Parameters[c,d]						van der Kelen Parameters[c]					
	k_1	k_2	c	d	t	p[e]	k_1	k_2	c	d	t	p	k_1	k_2	c	d	t	p
$Mn(CO)_5Cl$	1624	1751	21.3	23.1	45.2	0.215	1622	1750	22.2	22.2	44.5	0.215	1624	1750	21.1	23.8	44.9	0.222
$Mn(CO)_5Br$	1635	1741	18.6	30.5	43.2	0.327	1626	1741	22.1	22.1	44.2	0.235	1628	1742	20.6	24.2	44.8	0.241
$Mn(CO)_5I$	1638	1729	18.1	28.6	41.8	0.339	1630	1729	21.3	21.3	42.7	0.267	1632	1729	19.4	23.9	43.4	0.267
$Mn(CO)_5CH_3$	1622	1681	24.3	31.0	47.4	0.400	1613	1684	25.1	25.1	50.1	0.293	1610	1683	26.8	22.8	49.6	0.259
$Mn(CO)_5D$[f]	1647	1688	25.7	30.4	48.5	0.432	1640	1681	25.8	25.8	51.7	0.335	1637	1690	27.3	24.0	51.3	0.302
$Fe(CO)_4Br_2$	1762	1829	22.0	9.8[g]	33.7	0.248[g]	1761	1824	23.9	23.9	47.8	0.618	1775	1809	38.7	16.5	33.1	1.176
$Fe(CO)_4I_2$	1745	1786	28.0	16.0	30.0	0.746	1734	1792	19.3	19.3	38.6	0.499	1741	1784	27.1	15.4	30.8	0.667

a) Nm^{-1}. Notation as in Figs. (1, 2).
b) Based on ^{13}CO data. Results for species $Mn(CO)_5X$ from (27, 28); results for $Fe(CO)_4X_2$ from (81).
c) Recalculated for this work from all $-^{12}CO$ data of (27, 28, 81).
d) Calculated for $Mn(CO)_5X$ by ignoring B_2 mode; for $Fe(CO)_4X_2$ by using sum rule for A_1 eigenvalues (see text).
e) Mixing parameter, given by $2d/[k(2) + t + 2c - k(1)]$ for $Mn(CO)_5X$, and by $2d/[k(2) + t - k(1) - c]$ for $Fe(CO)_4X_2$.
f) Values for $Mn(CO)_5H$ distorted by coupling between C—O and Mn—H stretching motions.
g) This result appears anomalously low.

VI. References

1. *Braterman, P. S.:* Struct. Bonding *10*, 57 (1972).
2. a) *Adams, D. M.:* Metal-ligand and related vibrations. London: Edward Arnold 1967;
   b) *Nakamoto, K.:* Infrared spectra of inorganic and Co-ordination compounds, 2nd edit. New York: John Wiley 1970;
   c) *Braterman, P. S.:* Metal carbonyl spectra. London: Academic Press 1975;
   d) *Dobson, G. R., Stolz, I. W., Sheline, R. K.:* Advan. Inorg. Chem. Radiochem. *8*, 1 (1966);
   e) *Haines, L. M., Stiddard, M. H. B.:* Advan. Inorg. Chem. Radiochem. *12*, 53 (1969).
3. *Wilson, E. B., jr., Decius, J. C., Cross, P. R.:* Molecular vibrations. New York: McGraw-Hill 1955.
4. *Miller, J. R.:* J. Chem. Soc. (A) *1971*, 1855.
5. a) *Ottesen, D. K., Gray, H. B., Jones, L. H., Goldblatt, M.:* Inorg. Chem. *12*, 1051 (1973);
   b) *Jones, L. H.:* Inorg. Chem. *7*, 1681 (1968).
6. *Jones, L. H.:* Advances in the chemistry of the coordination compounds (Kirschner, S., ed.), p. 398. New York: MacMillan 1961.
7. *Cotton, F. A., Kraihanzel, C. S.:* J. Am. Chem. Soc. *84*, 4432 (1962).
8. *Kraihanzel, C. S., Cotton, F. A.:* Inorg. Chem. *2*, 533 (1963).
9. *Jones, L. H., McDowell, R. S., Goldblatt, M.:* Inorg. Chem. *8*, 2349 (1969).
10. *Buttery, H. J., Keeling, G., Kettle, S. F. A., Paul, I., Stamper, P. J.:* J. Chem. Soc. (A) *1969*, 2077.
11. *Bor, G., Sbrignadello, G.:* J. C. S. Dalton *1974*, 440.
12. *Braterman, P. S., Thompson, D. T.:* J. Chem. Soc. (A) *1968*, 1454.
13. *Braterman, P. S.:* J. Chem. Soc. (A) *1968*, 2907.
14. *El-Sayed, M. A., Kaesz, H. D.:* J. Mol. Spectry. *9*, 310 (1962).
15. *Haas, H., Sheline, R. K.:* J. Chem. Phys. *47*, 2996 (1967).
16. *Craig, D. P., Walsh, J. R.:* J. Chem. Soc. *1958*, 1603 and references therein.
17. *Bigorgne, M.:* J. Organometal. Chem. *1*, 101 (1963).
18. *Bouquet, G., Bigorgne, M.:* J. Mol. Struct. *1*, 211 (1967).
19. *Bigorgne, M., Poilblanc, R., Pańkowski, M.:* Spectrochim. Acta *26A*, 1217 (1970).
20. a) *Lewis, J., Manning, A. R., Miller, J. R., Ware, M. J., Nyman, F.:* Nature *207*, 142 (1965);
    b) *Cotton, F. A., Wing, R. M.:* Inorg. Chem. *4*, 1328 (1965).
21. *Wilkes, G. R.,* cited by *Wei, C. H.:* Inorg. Chem. *8*, 2384 (1969).
22. *Cariati, F., Valenti, V., Zerbi, G.:* Inorg. Chim. Acta *3*, 378 (1969).
23. *Edgar, K., Lewis, J., Manning, A. R., Miller, J. R.:* J. Chem. Soc. (A) *1968*, 1217.
24. *Bor, G.:* Inorg. Chim. Acta *1*, 81 (1967).
25. *Jernigan, R. T., Brown, R. A., Dobson, G. R.:* J. Coord. Chem. *2*, 47 (1972).
26. *Bor, G.:* Inorg. Chim. Acta *3*, 191 (1969).
27. *Kaesz, H. D., Bau, R., Hendrickson, D., Smith, J. M.:* J. Am. Chem. Soc. *89*, 2844 (1967).
28. *Braterman, P. S., Harill, R. W., Kaesz, H. D.:* J. Am. Chem. Soc. *89*, 2851 (1967).
29. *Poilblanc, R., Bigorgne, M.:* J. Organometal. Chem. *5*, 93 (1966).
30. *Dalton, J., Paul, I., Smith, G., Stone, F. G. A.:* J. Chem. Soc. (A) *1968*, 1195.
31. *Delbeke, F. T., Claeys, E. G., Van der Kelen, G. P., de Caluwe, R. M.:* J. Organometal. Chem. *23*, 497 (1970).
32. *Delbeke, F. T., Claeys, E. G., de Caluwe, R. M., Van der Kelen, G. P.:* J. Organometal. Chem. *23*, 505 (1970).
33. *Bouquet, G., Bigorgne, M.:* Spectrochim. Acta *27A*, 139 (1971).
34. *Jones, L. H., McDowell, R. S., Goldblatt, M.:* J. Chem. Phys. *48*, 2663 (1968).
35. *Redlich, O.:* Z. Physik. Chem. (B) *28*, 371 (1935). — *Teller, E.:* quoted by *Angus, W. R., Bailey, C. R., Hale, J. B., Ingold, C. K., Leckie, A. H., Raisin, C. G., Thompson, J. W.,* and *Wilson, C. L.:* J. Chem. Soc. *1936*, 971.
36. *Noack, K.:* Helv. Chim. Acta *47*, 1555 (1964).
37. *Huggins, D. K., Kaesz, H. D.:* J. Am. Chem. Soc. *86*, 2734 (1964).
38. *Noack, K., Ruch, M.:* J. Organometal. Chem. *17*, 309 (1969).
39. *Noack, K.:* private communication.
40. *Bor, G.:* Inorg. Chim. Acta *3*, 195 (1969); Appendix to Ref. (*26*).

41. *Bor, G., Johnson, B. F. G., Lewis, J., Robinson, P. W.:* J. Chem. Soc. (A) *1971*, 696.
42. *Perutz, R. N., Turner, J. J.;* Inorg. Chem. *14*, 262 (1975).
43. *Hertzberg, G.:* Spectra of diatomic molecules. Princeton: Van Nostrand 1950.
44. *Smith, J. M., Jones, L. H.:* J. Mol. Spectry. *20*, 248 (1966).
45. *Miller, J. R.:* private communication.
46. *Graham, M. A., Poliakoff, M., Turner, J. J.:* J. Chem. Soc. (A) *1971*, 2939.
47. *Hales, L. A. W., Irving, R. J.:* Spectrochim. Acta *23A*, 2981 (1967).
48. *Fenske, R. F., DeKock, R. L.:* Inorg. Chem. *9*, 1053 (1970).
49. *Caulton, K. G., Fenske, R. F.:* Inorg. Chem. *7*, 1273 (1968).
50. *Strohmeier, W.;* Fortschr. Chem. Forsch. *10*, 306 (1968). — *Werner, H.:* Angew. Chem. Intern. Ed. Engl. *7*, 930 (1968).
51. *Fischer, E. O., Maasböl, A.:* Angew. Chem. Intern. Ed. Engl. *3*, 580 (1964).
52. *Black, J. D.:* Ph. D. Thesis, Glasgow, 1975.
53. *McClure, D. S.:* Electronic spectra of molecules and ions in crystals. New York: Academic Press 1959, and references therein.
54. *Buttery, H. J., Kettle, S. F. A., Kontik-Matecka, B. T.:* Spectrochim. Acta *28A*, 1571 (1972).
55. *Adams, D. M., Hooper, M. A., Squire, A.:* J. Chem. Soc. (A) *1971*, 71.
56. *Adams, D. M., Fernando, W. S., Hooper, M. A.:* J. C. S. Dalton *1973*, 2264.
57. *Kettle, S. F. A., Paul, I.:* Advan. Organometal. Chem. *10*, 199 (1972).
58. *Noack, K.:* Helv. Chim. Acta *45*, 1847 (1962).
59. *Ramsay, D. A.:* J. Am. Chem. Soc. *74*, 72 (1952).
60. *Bellamy, I. J.:* Infrared spectra of complex molecules 2nd edit. London: Methuen 1958.
61. *Darensbourg, D. J., Brown, T. L.:* Inorg. Chem. *7*, 959 (1968).
62. *Abel, E. W., McLean, R. A. N., Tyfield, S. P., Braterman, P. S., Walker, A. P.:* J. Mol. Spectry. *30*, 29 (1969).
63. *Beck, W., Nitzmann, R. E.:* Z. Naturforsch. *17B*, 577 (1962).
64. *Bigorgne, M., Benlian, D.:* Bull. Soc. Chim. France, *1967*, 4100. — *Benlian, D., Bigorgne, M.:* Bull. Soc. Chim. France *1967*, 4106.
65. *Braterman, P. S., Bau, R., Kaesz, H. D.:* Inorg. Chem. *6*, 2097 (1967).
66. *Poliakoff, M., Turner, J. J.:* J. C. S. Dalton *1974*, 2276.
67. *Keeling, G., Kettle, S. F. A., Paul, I.:* J. Chem. Soc. (A) *1971*, 3143.
68. Ref. (*20*b), p. 1131.
69. *Wing, R. M., Crocker, D. C.:* Inorg. Chem. *6*, 289 (1967).
70. *Lewis, J., Manning, A. R., Miller, J. R.:* J. Chem. Soc. (A) *1966*, 845.
71. Ref. (*67*), p. 962.
72. *Cotton, F. A.:* Inorg. Chem. *3*, 707 (1964).
73. *Cossee, P., Schachtschneider, H. J.:* J. Chem. Phys. *44*, 97 (1966).
74. *Becher, H. J., Adrian, A.:* J. Mol. Struct. *7*, 323 (1971).
75. *Hollenstein, H., Guenthard, Hs. H.:* Spectrochim. Acta *27A*, 2027 (1971).
76. *Dobson, G. R.:* Inorg. Chem. *4*, 1673 (1965).
77. *Hall, M. B., Fenske, R. F.:* Inorg. Chem. *11*, 1619 (1972).
77a. *Fenske, R. F.:* private communication.
78. *Graham, W. A. G.:* Inorg. Chem. *7*, 315 (1968).
79. *Brown, R. A., Dobson, G. R.:* Inorg. Chim. Acta *6*, 65 (1972).
80. *Appleton, T. G., Clark, H. C., Manzer, L. E.:* Coord. Chem. Revs. *10*, 335 (1973).
81. *Johnson, B. F. G., Lewis, J., Robinson, R. W., Miller, J. R.:* J. Chem. Soc. (A) *1968*, 1043.
82. *Dessy, R. E., Wieczorek, L.:* J. Am. Chem. Soc. *91*, 4963 (1969).
83. *Bor, G., Jung, G.:* Inorg. Chim. Acta *3*, 69 (1969).
84. *Poilblanc, R.:* Compt. Rend. Acad. Sci. Paris, Sér. C *256*, 4910 (1963).
85. *Adams, D. M.:* J. Chem. Soc. (A) *1969*, 87.

Received March 17, 1975

The Influence of Charge-Transfer and Rydberg States on the Luminescence Properties of Lanthanides and Actinides

G. Blasse

Physical Laboratory, Sorbonnelaan 4, Utrecht, The Netherlands

Table of Contents

Note added in proof

Relating to Section 3a (the Eu^{3+} ion) we should like to refer the reader to a recent review by *Peacock* in this series on the intensities of lanthanide $f-f$ transitions (*111*).

Relating to Section 6 we mention a very recent observation by *De Hair* (*112*) who found that the long-wavelength ultraviolet excitation band of the uranate emission in ordered perovskites shows vibrational fine structure. The relevant phonons can be assigned to the excited state of the uranate octahedron. This suggests very strongly that this excitation band must correspond to a c.t. from filled "gerade" molecular orbitals of the oxygen ions to the $5f$ orbitals of uranium. This observation does not influence the discussion given above.

G. Blasse

Summary

The influence of charge-transfer and rydberg ($f^{n-1}d$) states on the luminescence of lanthanides and actinides (more specifically hexavalent uranium) is demonstrated. Optical transitions between the ground state and these excited states, and their dependence on electron configuration and crystallographic surroundings of the complex involved, are discussed first. The influence of charge-transfer and $f^{n-1}d$ states on emission spectra is dealt with. It is shown that the characteristics of emission spectra are often very sensitive to the energetic position of these states. Even more drastic is their influence on the temperature quenching of these emissions. Especially the case of the Eu^{3+} ion in oxysulfides is reviewed extensively. Direct feeding of the ^5D levels of Eu^{3+} by charge-transfer states occurs, but the reverse process is also possible. The spectral position of transitons to charge-transfer and $f^{n-1}d$ states can influence energy transfer probabilities. Metal ion-metal ion charge transfer states are also of great importance in this field. Finally some luminescence properties of hexavalent uranium are discussed. It is shown that often this emission is due to an octahedral UO_6^{6-} group and not to the well-known uranyl (UO_2^{2+}) group. Charge-transfer states involving $5f$ and possibly $6d$ levels determine the dependence of the emission characteristics on the host lattice.

1. Introduction

Luminescence from lanthanide and actinide ions has been a well-known phenomenon for many years. Especially during the last fifteen years luminescence of this type has been investigated thoroughly. Many compounds with interesting luminescence properties have been found, part of which have found practical application (*e.g.* Y_2O_2S—Eu^{3+} in colour television tubes, YVO_4—Eu^{3+} in luminescent lamps). Many unexpected phenomena could be realized using the lanthanides (*e.g.* anti-Stokes emission and two-quantum emission). It is the purpose of this paper to illustrate the influence of charge-transfer and excited $f^{n-1}d$ states on the lanthanide and actinide luminescence. Up till a few years ago this influence has been underestimated. Recent research has shown that many properties of lanthanide $4f$—$4f$ emission are determined by the role that is played by the charge-transfer or $f^{n-1}d$ states. Since the lanthanide ions incorporated in suitable host lattices often are ideal physical model systems, it may be expected that a better understanding of this role may also be of great help in a further explanation of luminescent phenomena.

The organization of this paper is as follows. In Section 2 we will consider the charge-transfer (c.t.) and $4f^{n-1}5d$ states of the lanthanides and also their dependence on the surroundings of the lanthanide ion. Although this paper is written from the point of view of a solid-state chemist, the results should also be applicable to solutions. The study of the solid state, however, has the advantage that coordination and further surroundings of the lanthanides can be chosen and fixed.

In Section 3 we will consider the influence of c.t. and $4f^{n-1}5d$ states on the emission spectra of the lanthanides. Since they often determine the relevant transition probabilities, the emission colours and luminescent decay times are determined to an important extent by the spectral position of c.t. and $4f^{n-1}5d$ states.

In Section 4 it will be shown that the thermal quenching temperature of the lanthanide emission depends strongly on the c.t. or $4f^{n-1}5d$ state and that progress has been made to understand this phenomenon.

In Section 5 the influence of c.t. or $4f^{n-1}5d$ states on the probability for energy transfer between two luminescent centres is dealt with. This influence is sometimes unexpectedly large.

In Section 6 we will consider one of the actinides, viz. hexavalent uranium. Its luminescence has been studied extensively, but many problems remain still unresolved. The position of the c.t. bands appears to be very important.

Nowhere we have aimed at completeness of data. Our main goal is to illustrate the influence of c.t. and $4f^{n-1}5d$ states on the luminescence of lanthanide ions with pronounced examples.

2. Charge-transfer and $4f^{n-1} 5d$ States of the Lanthanides

Jørgensen (1) has been the first who assigned the broad and strong absorption bands in the spectra of the trivalent lanthanides to either charge-transfer or $4f \rightarrow 5d$ transitions. *Nugent* and co-workers *(2)* have reported transitions of this kind for the actinides. In Table 1 we give some illustrative examples.

Table 1[a])

First c.t. absorption bands for some Ln^{3+} and An^{3+} ions

Species	ν(kK)	ε_{max} $(M^{-1}cm^{-1})$	δ(kK)
$SmCl_6^{3-}$	43.1	930	2.3
$SmBr_6^{3-}$	35.0	1050	2.4
$EuCl_6^{3-}$	33.2	400	2.1
$TmCl_6^{3-}$	~46.8		
$YbCl_6^{3-}$	36.7	160	1.7
$CfCl_6^{3-}$	42.1	1550	1.9
$EsCl_6^{3-}$	38.1	~500	1.6

$4f \rightarrow 5d$ absorption bands for Ln^{3+} ions and $5f \rightarrow 6d$ absorption bands for An^{3+} ions

Species	ν(kK)	ε_{max} $(M^{-1}cm^{-1})$	δ(kK)
$CeCl_6^{3-}$	30.3	1600	0.8
$CeBr_6^{3-}$	29.15	1600	1.05
$PrCl_6^{3-}$	42.0		
$TbCl_6^{3-}$	36.8	30	1.2
	42.75	1500	0.7
UCl_6^{3-}	15.6	1400	
$AmCl_6^{3-}$	38.5	500	1.2
	41.0	15100	0.7
	43.7	7100	1.0
	46.7	5200	1.1

[a]) from a compilation in Ref. *(2)*. ν: wavenumber at maximum absorption; ε_{max}: molar absorptivity; δ: half-width.

As a general rule the c.t. bands shift to lower energies with increasing oxidation state, whereas Rydberg transitions (such as $4f \rightarrow 5d$ transitions) shift to higher energies. It may, therefore, be expected that the lowest absorption bands of the tetravalent lanthanide ions will be due c.t. transitions and those of the divalent lanthanide ions to $4f \rightarrow 5d$ transitions.

Let us consider the tetravalent lanthanide ions first. Because this valency is hard to realize, the number of investigations is restricted. *Jørgensen* and *Rittershaus (3)* described the diffuse reflection spectra of Pr^{4+} and Tb^{4+} in ThO_2 and Y_2O_3 and *Blasse* and co-workers those in ZrO_2 *(4)*. These spectra show strong absorption bands in the visible which were ascribed to c.t. transitions.

In ThO_2, however, they were situated at considerably lower energy than in ZrO_2, e.g. Tb^{4+} in ThO_2 at 20.3 kK and in ZrO_2 at 28.6 kK. These c.t. bands have been studied in more detail by *Hoefdraad* (5). He introduced Ce^{4+}, Pr^{4+} and Tb^{4+} in a number of oxidic host lattices where the coordination of the Ln^{4+} ions would be fixed, viz. either six- or eight-coordination. His results are as follows:

a) In six-coordination the position of these c.t. bands does not depend on the host lattice. The position of the c.t. band in $BaZrO_3$ is the same as in $SrZrO_3$, as in Ba_2ZrO_4, as in $BaMO_3(M=Ce, Hf, Th)$.

b). In eight-coordination, however, this position depends on the host lattice in such a way that the Ln^{4+}–O^{2-} distance influences the spectral position of the c.t. band. In Fig. 1 some of these spectra are given.

Hoefdraad was able to explain his results with a relatively simple model. Following *Schmidtke* (6) he assumed the sequence of ligand m.o.'s in an octahedron to be

$$a_{1g} < 1t_{1u} < t_{2g} < t_{2u} < t_{1g} = e_g < 2t_{1u} .$$

Since the e_g orbitals are stabilized by interaction with the d-orbitals of the central ion, he ascribed the first c.t. transition to the parity-allowed $t_{1g} \rightarrow f$. Since neither the t_{1g} nor the f level are strongly influenced by the bonding in the complex, the position of the c.t. transition is expected to be host-lattice independent as observed experimentally for six-coordination.

In passing we mention that the c.t. transitions of Ti^{4+} in these host lattices are crystal-structure dependent (7). These absorption bands must be ascribed to the $2t_{1u} \rightarrow d$ transition (8). Indeed these levels depend on the nature of the bonding

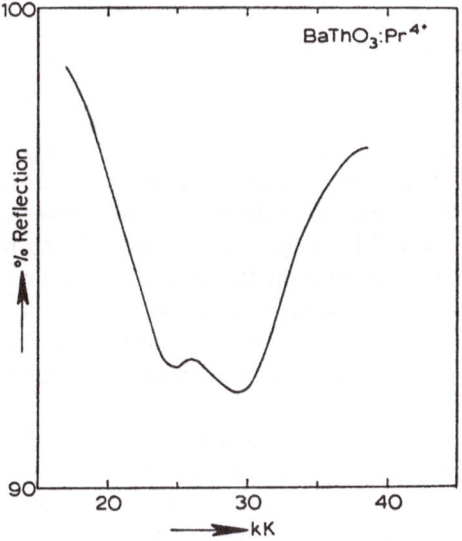

Fig. 1a. Diffuse reflection spectrum of $BaThO_3$-Pr^{4+} at room temperature (Pr^{4+} in six-coordination)

G. Blasse

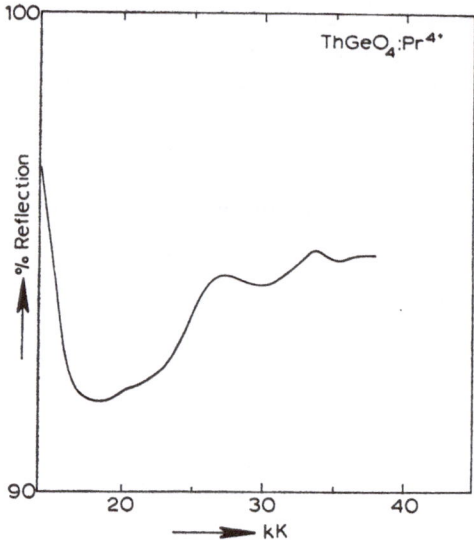

Fig. 1b. Diffuse reflection spectrum of $ThGeO_4$-Pr^{4+} at room temperature (Pr^{4+} in eight-coordination)

in the complex. Note that the case of Ti^{4+} and Ln^{4+} ions in zirconates like those mentioned is exceptional, because the absorption spectra of transition metal ions and lanthanide ions can be studied in a given crystal lattice where these ions occupy one and the same crystallographic site. This is seldom possible.

For cubic eight-coordination the sequence of ligand m.o's is

$$a_{1g} = e_g < 1t_{1u} = t_{2u} < 1t_{2g} < 2t_{1u} < t_{1g} = 2t_{2g} < e_u = a_{2u} \qquad (5)$$

The c.t. transitions of Ln^{4+} ions in eight-coordination are ascribed to the transition $2t_{2g} \rightarrow f$. Note that there are two ligand m.o's of t_{2g} symmetry. These are perturbed by the $t_{2g}(5d)$ metal ion orbital. As a consequence the position of the $2t_{2g}$ level depends on the energetic position and the extent of delocalization of the d-orbitals of the Ln^{4+} ion. This explains the dependence of the first c.t. band of Ln^{4+} ions in eight-coordination on the choice of the host lattice.

After this discussion of c.t. transitions on Ln^{4+} ions we now turn to the divalent lanthanides ions. Here the first allowed transitions in the spectra are $4f \rightarrow 5d$ transitions as expected. They have been studied in detail. We will here mention some relevant results.

The $4f \rightarrow 5d$ transitions of nearly all Ln^{2+} ions have been observed in CaF_2 (9). Luminescence from these transitions has been studied in detail for Eu^{2+} (10), Sm^{2+} (11) and Yb^{2+} (12). In good approximation these spectra can be ascribed as transitions between the $4f^n$ ground state and the d crystal-field components of the $4f^{n-1}d$ state. The influence of the surroundings on these transitions is twofold: the position of the centre of gravity of the $5d$ level is influenced by the nature of the surrounding ions or ligands [nephelauxetic effect, Ref. (13)],

48

Table 2. Absorption bands, crystal-field splitting and centre of gravity of the $5d$ level of Eu^{2+} in several host lattices (all values in kK)

Host lattice	Coordination Eu^{2+} ion	Absorption bands		Crystal field splitting	Centre $5d$ level	Ref.
CaF_2	cube	27.1	45.0	17.9	37.9	(14)
SrF_2	cube	27.9	43.5	15.6	37.3	(14)
BaF_2	cube	28.5	42.7	14.2	37.0	(14)
NaCl	octahedron	29.4	41.5	12.1	34.2	(15)
KCl	octahedron	29.1	40.1	11.0	33.5	(16)
		29.7	41.4	10.7	34.4	(15)
KBr	octahedron	29.0	40.2	11.2	33.5	(16)
		29.7	38.1	8.4	33.1	(15)
$BaZrO_3$	regular 12 coordination	25.2	38.2	13.0	33.0	(17)

but the crystal-field splitting of this $5d$ level depends also on the nature and arrangement of these ions. In Table 2 we have given some examples for the case of Eu^{2+}.

We now turn to the common valency of the lanthanides, viz. three. Here we find that depending on the number of f electrons in the ground state the first allowed transition may be either a c.t. transition or a $4f \rightarrow 5d$ transition. The stability of the half-filled and completely-filled shells serves as a starting point to predict which of the two transitions is to be expected for a special case: the c.t. transitions are at relatively low energy in the case of $Eu^{3+}(4f^6)$ and $Yb^{3+}(4f^{13})$, the $4f \rightarrow 5d$ transitions are at relatively low energy in the case of $Ce^{3+}(4f^1)$, $Pr^{3+}(4f^2)$, $Tb^{3+}(4f^8)$. Loh (18) has measured the lowest $4f \rightarrow 5d$ transitions of all trivalent lanthanides in CaF_2 (see Table 3).

Table 3. Lowest $4f \rightarrow 5d$ absorption band of Ln^{3+} ions in CaF_2 at room temperature (18)

Ln^{3+}	ν(kK)
Ce^{3+}	32.5
Pr^{3+}	45.6
Nd^{3+}	55.9
Sm^{3+}	59.5
Eu^{3+}	68.5
Gd^{3+}	> 78.0
Tb^{3+}	46.5
Dy^{3+}	58.9
Ho^{3+}	64.1
Er^{3+}	64.2
Tm^{3+}	64
Yb^{3+}	70.7
Lu^{3+}	> 80.0
Tb^{4+}	75.4

We will now try to account for the way in which these two types of transitions are influenced by the surrounding ligands.

a). c.t. transitions. There is ample evidence that for a given lanthanide ion the position of the c.t. transition is at lower energy if the surrounding ligands are more reducing (or less electronegative). This well-known fact will here not be discussed further. We note that there is a tendency to have the c.t. transition at lower energy, if the number of surrounding ligands is larger (19—21). Finally the dependence on the host lattice for a given coordination with the same kind of ligands follows from the work of *Hoefdraad* (5, 22). The latter is especially of importance in oxides. In Table 4 we have illustrated these rules for the case of Eu^{3+}.

Table 4. Position of the first c.t. band of Eu^{3+} in some oxides [after Refs. (21, 22)]

Compound	Coordination	Position c.t. band (kK)
$ScBO_3$-Eu^{3+}	6	43.0
$LiLuO_2$-Eu^{3+}	6	43.0
$NaYGeO_4$-Eu^{3+}	6	43.1
YBO_3-Eu^{3+}	6	42.7
Y_2O_3-Eu^{3+}	6	41.7
$NaGdO_2$-Eu^{3+}	6	41.1
$CaYBO_4$-Eu^{3+}	6	41.7
$ScPO_4$-Eu^{3+}	8	~48.0
YPO_4-Eu^{3+}	8	~45.0
$Y_3Ga_5O_{12}$-Eu^{3+}	8	42.5
$YTaO_4$-Eu^{3+}	8	40.8
$LaPO_4$-Eu^{3+}	8	37.0
$LaTaO_4$-Eu^{3+}	8	36.0
$GdGaO_3$-Eu^{3+}	12	40.5
$LaAlO_3$-Eu^{3+}	12	32.3
$SrLaLiWO_6$-Eu^{3+}	12	30.5

b) $4f \rightarrow 5d$ transitions. As mentioned above the spectral position of these transitions in determined mainly by the nephelauxetic effect and the crystal field effecting the $5d$ level. Some examples are given in Table 5 for Ce^{3+}.

Finally we note some other properties of these allowed transitions of the lanthanide ions. From Table 1 it becomes clear that in general the $4f$—$5d$ bands have a smaller band width than the c.t. transitions, typical values being 1000 and 2000 cm^{-1}, respectively. In this connection it is interesting to find that at low temperatures the $4f \rightarrow 5d$ absorption and emission bands often show a distinct and extended vibrational fine structure [Ce^{3+} (25), Tb^{3+} (25), Eu^{2+} (14, 26), Yb^{2+} (27)], whereas c.t. transitions do not. From this it seems probable that in the excited c.t. state the interaction between the lanthanide ion and its surroundings is stronger than in the excited $4f^{n-1}5d$ state. This is not unexpected. As far

Table 5. The 5d levels and cubic crystal-field splitting (Δ) for Ce^{3+} in several host lattices. All values in kK (23)

Composition	Coordination of Ce^{3+}	5d levels (observed)[a]	Deduced cubic levels	Cubic crystal-field splitting (Δ)	Spherical 5d
$Y_3Al_5O_{12}$-Ce	Distorted cube (D_2)	22.0 29.4|37 44	25.7 (e_g), ~40(t_{2g})	~14	~34.5
$YAl_3B_4O_{12}$-Ce	Trigonal prism (D_{3h})	31.0 36.6 39.2	...	~6.5[c]	35.4
$ScBO_3$-Ce	Distorted octahedron (D_{3d})	28.0 31.2|36.1 38.5	~29.5(t_{2g}), 37.3(e_g)	~8.0	~32.5
$Sc_2Si_2O_7$-Ce	Distorted octahedron (C_2)	29.0 33.3|43.5	~31.0(t_{2g}), ~43.5(e_g)	~12.5	~35.0
SrF_2-Ce[b]	Distorted cube	33.6 48.8|50.3 53.4	41.2(e_g), 52.4(t_{2g})	11.2	48.0
Free ion[b]	...	...	...	...	51.0

a) Vertical bar indicates separation between lower and higher cubic levels.
b) After Ref. (24).
c) Value of the octahedral crystal-field splitting derived from the position of the levels for Ce^{3+} in prismatic coordination using a point-charge model.

51

as we are aware nobody has ever reported vibrational fine structure for the c.t. transitions. Note also that luminescence from c.t. states has not been observed for the lanthanides, whereas luminescence from $4f^{n-1}5d$ states is quite common (Ce^{3+}, Eu^{2+}).

In later sections of this paper the influence of other c.t. transitions will show to be of importance too. These may be of two types, viz. ligand-to-central-metal transitions in the surrounding groups and metal-to-metal transitions. The former type has been discussed and reviewed extensively (*13, 28*). An example of relevance within this type is the UV absorption transition in the vanadate (VO_4^{3-}) group Examples of the second type are absorption bands ascribed to electron transfer between Pb^{2+} and W^{6+} (*29*) and Tb^{3+} and V^{5+} (*30*) in oxides.

With the present knowledge we now turn to our proper subject, viz. the influence of these c.t. and $4f^{n-1}5d$ states on the luminescence properties of lanthanides.

3. Influence of Charge-transfer and $4f^{n-1}5d$ States on the Emission Spectra of Lanthanides

In this section we will present examples of the influence of c.t. and $4f^{n-1}5d$ states on the emission spectra of some lanthanides.

a) The Eu^{3+} Ion

The emission of the Eu^{3+} ion consists of transitions from the ^5D manifold (mainly from the $^5D_0$ level) to the ^7F manifold. There is one report of c.t. emission in the case of Eu^{3+} (in aqueous solutions) (37), but this has not been confirmed by other investigations. If we restrict ourselves to $^5D_0 \rightarrow ^7F$ emission, the selection rules (32) allow in first approximation the following transitions:

a) magnetic dipole emission: $^5D_0 \rightarrow ^7F_1$ (insensitive of crystallographic surroundings).

b) forced electric dipole emission: $^5D_0 \rightarrow ^7F_{0,2,4,6}$ (if inversion symmetry at the Eu^{3+} site is lacking).

Forced electric dipole emission occurs if it is possible to mix even functions into the uneven $4f$ functions, so that the parity selection rule is relaxed. It is usually assumed that this occurs by $4f$–$5d$ mixing. For Eu^{3+}, however, the $4f^5 5d$ state is at very high energy (see Table 3). Since the electric-dipole emission dominates for Eu^{3+} on sites without inversion symmetry, it seems obvious to assume that another state is used to relax the parity selection rule. This must occur by mixing the $4f^6$ configuration with the levels of opposite parity of the c.t. state.

This is nicely confirmed by a study of some Eu^{3+}-activated phosphates and vanadates with zircon structure (33). In Table 6 we give the observed ratio of

Table 6. Ratio of electric to magnetic dipole emission intensity and position of the lowest excitation band for some Eu^{3+}-activated ABO_4 compounds [after Ref. (33)]

Compound	Excitation band (kK)	Ratio $\dfrac{E.D.}{M.D.}$
$ScPO_4$-Eu^{3+}	~48.0	1.6
YPO_4-Eu^{3+}	~45.0	2.3
YVO_4-Eu^{3+}	31.2	6.5
$ScVO_4$-Eu^{3+}	29.8	8.4

electric to magnetic dipole emission of the Eu^{3+} luminescence in these hosts. We see that there is a correlation between the position of the lowest excitation (and absorption) band of these materials and the intensity ratio. This absorption band is a c.t. transition in which either europium or vanadium or both are involved. It has, therefore, been proposed that the parity-forbidden $4f \rightarrow 4f$ transitions of the Eu^{3+} ion borrow intensity from the lowest strong absorption band (either host lattice absorption or charge-transfer absorption within the Eu^{3+} center) and not from the $4f \rightarrow 5d$ absorption band. This explains the results of Table 6 reasonably well, since the borrowing will be more effective if the strong absorption band is at lower energies. In conclusion we find that for intense forced electric-dipole emission from Eu^{3+} two conditions must be fulfilled, viz. absence of inversion symmetry at the Eu^{3+} crystallographic site and c.t. transitions at low energies.

It is also interesting to compare similar results for Eu^{3+} in trigonal prismatic coordination (30). The ratio of electric to magnetic dipole Eu^{3+} emission is 1.2 for YF_3 : Eu, 4.0 for $GdTiSbO_6$: Eu and 8.0 for $YAl_3B_4O_{12}$: Eu. Note the much lower ratio for the fluoride than for the oxides which must be due to the difference in position of the c.t. band (in fluorides much higher than in oxides).

Similar results have been reported for Eu^{3+} in glasses (34): germanate glasses where the Eu^{3+} c.t. band is situated at 38462 cm^{-1} show a more intense forced-electric-dipole emission than phosphate glasses, where the c.t. band lies at 49020 cm^{-1}. These examples illustrate the influence of the c.t. state upon the Eu^{3+} $4f \rightarrow 4f$ emission.

There is ample evidence that the influence of $4f^{n-1}5d$ states on the $4f$—$4f$ emission spectra is not as pronounced as the influence of the c.t. state [Tb^{3+} (30), Tm^{3+} and Er^{3+} (35)]. For the Tb^{3+} ion, however, this is only seeming, because the emission spectrum is not strongly influenced as in the case of Eu^{3+}, but the life time τ of the luminescent $^5D_4$ level shows the influence of a relatively low $4f^75d$ state. This can be illustrated on the following examples where the Tb^{3+} ion occupies a crystallographic position without inversion symmetry. $YTaO_4$-Tb and $LaTaO_4$-Tb have $\tau \simeq 1$ msec with the $4f \rightarrow 5d$ transition at 37 kK, YBO_3-Tb and $Sr_3(PO_4)_2$-Tb have $\tau = 10$ and 9 msec, respectively and the $4f \rightarrow 5d$ transition at about 42 kK (36—38). For comparison, $InBO_3$-Tb^{3+} with Tb^{3+} on a site with inversion symmetry has $\tau \simeq 30$ msec (37).

In connection with the Eu^{3+} ion we finally call attention to the influence of the position of the c.t. state on the so-called hypersensitive transitions. These are transitions that are remarkably sensitive to the immediate environment of the lanthanide ion (39). They conform the selection rules for electric quadrupole radiation, $\Delta J = 2$. As such, the $^5D_0 \rightarrow {}^7F_2$ emission line of Eu^{3+} is hypersensitive as has also been shown for the solid state (40). In YF_3-Eu^{3+}, however, this transition is not hypersensitive, since the dominating forced electric-dipole emission transition is $^5D_0 \rightarrow {}^7F_4$ (see Fig. 2). In oxides, however, $^5D_0 \rightarrow {}^7F_2$ is hypersensitive and dominates the emission, if Eu^{3+} is not at a site with inversion symmetry. The application of Eu^{3+}-activated phosphors in television tubes and luminescent lamps is for an important part based upon this hypersensitivity. If the c.t. band of Eu^{3+} in oxides is at relatively high energy (say 40 kK), the $^5D_0 \rightarrow {}^7F_2$ emission may dominate. *Judd* (41) has proposed that linear crystal-field terms are of importance to explain the hypersensitivity, but the $^5D_0 \rightarrow {}^7F_2$ transition shows also hypersensitivity, if a linear crystal-field term vanishes in view of the site symmetry [e.g. in $NaGdO_2$-Eu^{3+} with D_{2d} site symmetry (42)]. It is striking however, that in those host lattices, where the Eu^{3+} ion occupies a site allowing a linear crystal-field term and the Eu^{3+} c.t. band is at relatively low energies, the $^5D_0 \rightarrow {}^7F_0$ emission is relatively strong (see Table 7). It is tempting to conclude that the combined presence of a linear crystal field term and a low-lying c.t. band contributes considerably to the intensity of the $^5D_0 \rightarrow {}^7F_{0,2}$ transitions. Hypersensitivity may, therefore, not be due to one single mechanism, but depending on

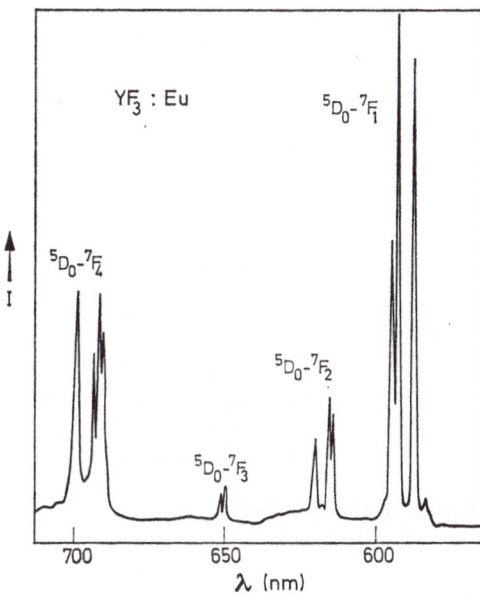

Fig. 2. Spectral energy distribution of the emission of YF_3-Eu (5%); 366 nm excitation (from Ref. (30))

Table 7. Intensity ratio of $^5D_0 \rightarrow {}^7F_0$ emission to $^5D_0 \rightarrow {}^7F_1$ emission for some Eu^{3+}-activated oxides (43)

Compound	Crystal structure	$\dfrac{^5D_0 \rightarrow {}^7F_0}{^5D_0 \rightarrow {}^7F_1}$	Maximum lowest excitation band (kK)
$Sr_2TiO_4:Eu^{3+},Na$	K_2NiF_4	1.65	~30.0
$SrLaAlO_4:Eu^{3+}$	K_2NiF_4	0.09	36.0
$Gd_2WO_6:Eu^{3+}$	Bi_2NbO_5F	0.68	32.0
$Ba_3Gd_2WO_9:Eu^{3+}$	perovskite	0.72	30.8
$Sr_2LaAlO_5:Eu^{3+}$	Cs_3CoCl_5	0.16	33.0
$LiGdO_2:Eu^{3+}$	$HAlO_2$	0.08	41.0

the site symmetry of the lanthanide ion and the position of the c.t. band to a forced electric-dipole mechanism via a linear term (42) or to a pseudoquadrupole mechanism (39) or, eventually, to a dynamic-coupling mechanism (44).

b) The Pr^{3+} Ion

The Pr^{3+} ion shows a number of different emissions depending on the host lattice in which it is incorporated, viz. red (from the $^1D_2$ level), green (from the $^3P_0$ level), blue (from the $^1S_0$ level) and ultraviolet (from the $4f5d$ state). The energy

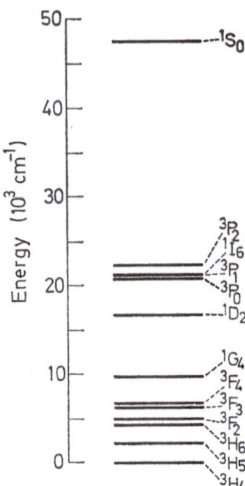

Fig. 3. Energy level diagram of the Pr^{3+} ion

level scheme of Pr^{3+} is shown in Fig. 3. The excited $4f5d$ state is one of the important factors that determine which of the emissions it to be expected.

In some fluorides (*e.g.* YF$_3$, LaF$_3$, NaYF$_4$) the lowest crystal-field component of the $4f5d$ state of Pr^{3+} is situated above the $^1S_0$ level. Excitation with short-wavelength ultraviolet radiation (*e.g.* 185 nm) or cathode-rays excites the Pr^{3+} ion from the $^3H_4$ ground state into the $4f5d$ level, from where it decays radiationless to the $^1S_0$ level (see Fig. 2). From here the Pr^{3+} ion returns to the ground state by two-photon luminescence (*45—47*): the emission spectrum contains a group of transitions in the blue and another group in the green and in the red. The latter is ascribed to emission from the $^3P_0$ level (*45*). The former is ascribed to the $^1S_0 \rightarrow \ ^3P_2$ transition by *Sommerdijk et al.* by comparison with data on the Pr^{3+} energy level scheme (*46*), and to the $^1S_0 \rightarrow \ ^1I_6$ transition by *Piper et al.* using a calculation of transition probabilities by the Judd-Ofelt theory (*45*). For YF$_3$ they find the parameter Ω_6 to be much larger than Ω_2 and Ω_4 which may also be the reason for the large intensity of the $^5D_0 \rightarrow \ ^7F_4$ emission of Eu^{3+} in YF$_3$ (see Fig. 1). For the present discussion it is not really important which of the two assignments is the correct one.

If, however, the lowest $4f5d$ state is below the $^1S_0$ level, two-photon luminescence is no longer observed. In a number of host lattices luminescence from this $4f5d$ state has been observed, e.g. LiYF$_4$, KYF$_4$, BaYF$_5$, YPO$_4$, Y$_2$(SO$_4$)$_3$ (*45*) and Y$_3$Al$_5$O$_{12}$ (*48*). CaF$_2$-Pr^{3+} is reported as a border-line case in Ref. (*45*), whereas *Sommerdijk et al.* did not observe two-photon luminescence for this material (*47*).

If the $4f5d$ levels are situated at still lower energy, no $5d \rightarrow 4f$ emission is observable. In stead emission from the $^3P_0$ level occurs. *Weber* (*48*) has studied in Y$_3$Al$_5$O$_{12}$-Pr^{3+} the nonradiative decay from the luminescent $4f5d$ level of Pr^{3+} to the $^3P_{0,1,2}$ and $^1I_6$ level. For temperatures below 250 K the decay time of the $5d \rightarrow 4f$ luminescence is constant and amounts to about 2.10^{-8} sec (as is to be expected for an allowed electric-dipole transition). Above 250 K the life-time of this $4f5d$ level decreases rapidly due to nonradiative decay to the $4f^2$ levels.

This is the situation in oxides where excitation into the $4f5d$ levels of Pr^{3+} is followed by emission from the $^3P_0$ level. In many cases, however, emission from $^1D_2$ occurs too. In tungstates, vanadates, niobates and related compounds the $^1D_2$ emission even dominates. Two models that are closely related have been proposed to explain effective $^3P_0 \rightarrow {}^1D_2$ relaxation:

a) *Reut* and *Ryskin* (49) have proposed a virtual recharge mechanism. This is shown schematically in the configuration coordinate diagram in Fig. 4. It is assumed that the charge-transfer state $Pr^{4+} + V^{4+}$ (if we take a vanadate lattice like YVO_4 as example) has a considerably larger (or smaller) equilibrium distance than the $Pr^{3+}(4f^2)$-V^{5+} states. Although this c.t. state is found above the $^3P_0$ level in absorption spectra, it facilitates radiationless decay from $^3P_0$ to $^1D_2$.

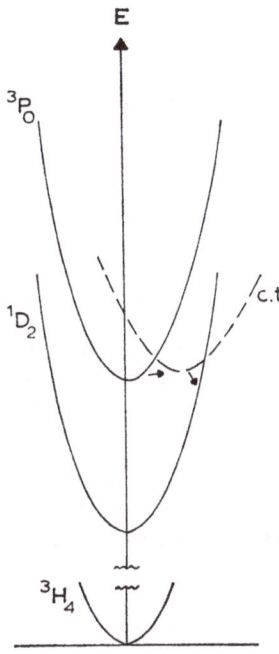

Fig. 4. Configuration coordinate diagram of Pr^{3+} showing radiationless decay from the $^3P_0$ to the $^1D_2$ level via a c.t. state (virtual recharge mechanism). The relevant $4f^2$ configuration levels have been drawn only. Note break in energy scale

b) *Hoefdraad* (50) has argued that the state with larger or smaller equilibrium distance may be a $4f5d$ level of Pr^{3+} itself. This can be the case, if it is at low enough energy. This model was illustrated by the Pr^{3+} emission in two calcium zirconates (the perovskite $CaZrO_3$ and the fluorite phase calcium-modified ZrO_2). In the fluorite phase the lowest $4f5d$ level is at about 34 kK and the emission occurs in equal amounts from $^3P_0$ and $^1D_2$. In the perovskite the lowest $4f5d$ level is at

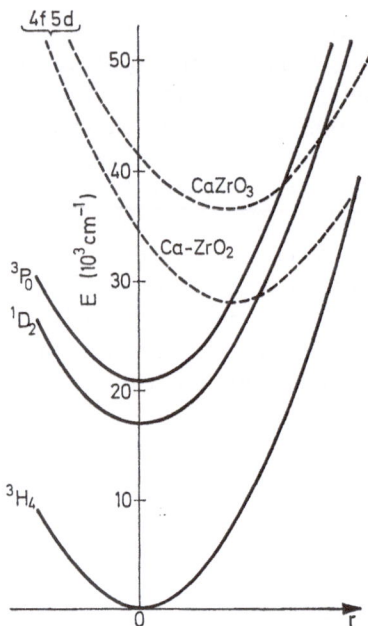

Fig. 5. Configuration coordinate diagram of Pr^{3+}. Drawn parabolas refer to levels of the $4f^2$ configuration, the upper broken parabola to the lowest level of the $4f5d$ configuration of Pr^{3+} in $CaZrO_3$ and the lower broken parabola to the lowest $4f5d$ level of Pr^{3+} in Ca-stabilized ZrO_2 (schematically)

41 kK and the emission contains practically only $^3P_0$ transitions. This model is drawn schematically in Fig. 5. It will be obvious that Δr, the difference between the equilibrium positions of the ground state and the excited $4f5d$ state will determine to a great extent which emission will be observed.

This shows the enormous influence of the $4f5d$ state of the Pr^{3+} ion on its emission characteristics. Two properties of the lowest $4f5d$ level are of importance, viz. its energetic position and the value of Δr. Later we will find that this statement is valid for more cases.

c) The Eu²⁺ Ion

Up till some years ago the Eu^{2+} ion was known as a broad band $4f^65d \rightarrow 4f^7$ emitter (10, 17). The ground state is $^8S(4f^7)$ and the lowest excited state $4f^65d$. Some years ago, however, sharp line emission for Eu^{2+} has been reported (51, 52). This means that the ^6P_J states of the $4f^7$ configuration are situated below the lowest $4f^65d$ level, so that the emission occurs within the $4f^7$ configuration. The position of the lowest $4f^65d$ level relative to the ^6P_J states determines, whether the Eu^{2+} ion will show narrow-line or broad-band emission. Narrow-line emission for Eu^{2+} is expected in lattices where the centre of gravity of the $4f^65d$ level is at high energy, the crystal-field is weak, and the cohesion energy is high (so that Δr, and consequently the Stokes shift of the broad-band emission, is small) (53). This is in qualitative agreement with the experimental results: narrow-line emission in many fluorides and also in strongly-bound oxides: $BaAlF_5$ (51, 52), $SrAlF_5$ (51, 52), $BaMg(SO_4)_2$ (26), and $SrBe_2Si_2O_7$ (54). The occurrence of line emission in compounds MeFX (Me = Sr, Ba and X = Cl, Br) (55, 56) is rather unexpected in view of the conditions mentioned above. These compounds have highly anisotropic crystal lattices. This is also the case for $SrAl_{12}O_{19}$-Eu^{2+} with line emission, whereas the analogous Ca and Ba compounds show band emission (57). The more intense line emission is in fact observed for the compounds with a relatively small Stokes shift of the band emission.

An interesting study in this connection is the work by *Ryan et al.* (26). They studied emission and excitation spectra of Eu^{2+} in $CaSO_4$ and $BaMg(SO_4)_2$ at 1.8 K. For $CaSO_4$-Eu^{2+} the emission (with decay time 0.4 μsec) is of the $5d \rightarrow 4f$ type and consists of a zero-phonon line followed by a large number of phonon replicas (due to density of states peaks in the normal lattice modes of $CaSO_4$). The excitation spectrum of this Eu^{2+} emission consists of 56 lines which are ascribed to purely electronic transitions to the levels of the $4f^6(^7F_J)5d(e_g)$ system, split by strong exchange interaction between the $4f$ and $5d$ electrons.

In $BaMg(SO_4)_2$-Eu^{2+}, however, the emission is of the $4f \rightarrow 4f$ type (decay time 3.5 msec) and consists of a zero-phonon line (the $^6P_{7/2} \rightarrow {}^8S_{7/2}$ transition) with a large number of phonon replicas at lower energy. The excitation band contains seven narrow bands ascribed to the seven ^7F_J levels of the Eu^{3+} core of the excited $4f^65d$ state. Obviously the exchange interaction between the $4f$ and $5d$ electrons is much smaller in $BaMg(SO_4)_2$ than in $CaSO_4$, which means that the $5d$ electron of the Eu^{2+} ion is stronger localized in $CaSO_4$ than in $BaMg(SO_4)_2$. Strong exchange interaction depresses the $4f^65d$ levels and suppresses, therefore, the sharp line emission.

Recently we have found that $BaMg(SO_4)_2$ has the $KAl(SO_4)_2$ structure which is an ordered variant of the NiAs structure. This means that layers of sulfate groups are surrounded by a barium layer on one side and a magnesium layer on the other side. The sulfate ions are therefore polarized towards the magnesium layer. As a consequence the Ba^{2+} ions will show a relatively high degree of covalent bonding and the Eu^{2+} ion (on the barium sites) a relatively high degree of delocalization of its excited $5d$ electron.

Finally we mention the emission of the $Sm^{2+}(4f^6)$ ion where a situation analogous so that of Eu^{2+} occurs. In Sm^{2+} the $^5D_0(4f^6)$ level and the lowest $4f^55d$ level are at about the same energy. If $^5D_0$ is lower, $4f \rightarrow 4f$ emission occurs, if the $4f^55d$ level is lower, $5d \rightarrow 4f$ emission occurs. *Wood* and *Kaiser* (11) have studied Sm^{2+} in CaF_2, SrF_2 and BaF_2. The emission spectrum of Sm^{2+} in CaF_2 is different from the spectra of Sm^{2+} in the other compounds: in CaF_2 we have $5d \rightarrow 4f$ emission, in SrF_2 and BaF_2 $4f \rightarrow 4f$ emission. Probably the larger crystal field in the calcium compound (with too small sites for the Sm^{2+} ion) is responsible for the variation in the emission along this series.

d) The Gd^{3+} Ion

The phenomena discussed up till now in this section can be described by considering energy levels of one and the same ion (except for the virtual recharge mechanism). Charge-transfer states of neighbouring ions or groups of ions can also influence the luminescence of lanthanide ions. This will be illustrated on the series $(Y,Gd)TaO_4$, $(Y,Gd)NbO_4$ and $(Y,Gd)VO_4$. The first excited level of the Gd^{3+} $(4f^7)$ ion is $^6P_{7/2}$ and the emission $^6P_{7/2} \rightarrow {}^8S$ is to be expected at about 313 nm. If $(Y,Gd)TaO_4$ is excited into the tantalate group ($\lambda < 225$ nm), efficient energy transfer from the tantalate group to Gd^{3+} occurs and the total emission contains a considerable amount of Gd^{3+} line emission (38). If $(Y,Gd)VO_4$ is excited into the vanadate group ($\lambda < 330$ nm), vanadate emission occurs, but no Gd^{3+} emission. After relaxation the vanadate group has reached such a low energy ($\simeq 400$ nm emission) that transfer to the Gd^{3+} ion is impossible. The situation for $(Y,Gd)NbO_4$ is in between these two extremes: there is some Gd^{3+} line emission, but much more broad band niobate emission. The occurrence of Gd^{3+} emission depends on the energetic position of the c.t. state of the host lattice groups. For the tantalate this state is high enough for Gd^{3+} emission to occur, in the vanadate it is so low that no Gd^{3+} emission is possible.

4. Influence of Charge-transfer and $4f^{n-1}5d$ States on the Temperature Quenching of Lanthanide Luminescence

During recent years it has become clear that the temperature quenching of lanthanide luminescence is determined by the properties of the c.t. or $4f^{n-1}5d$ state. Let us first mention shortly the mechanisms that have been proposed for temperature quenching of lanthanide ion emission.

Temperature quenching of broad band emission is usually explained by a simple configuration coordinate model consisting of two parabolas that have been shifted with regard to each other (Fig. 6). This is called the Mott-Seitz model. Nonradiative return from the excited to the ground state is possible via the parabola crossover. Its rate can be described with an activation-energy formula $P_{nr} = C\,e^{-\Delta E/kT}$, where C is a constant of the order of 10^{13} sec^{-1} and ΔE is the activation energy to the crossover. For details we refer to the literature (58, 59).

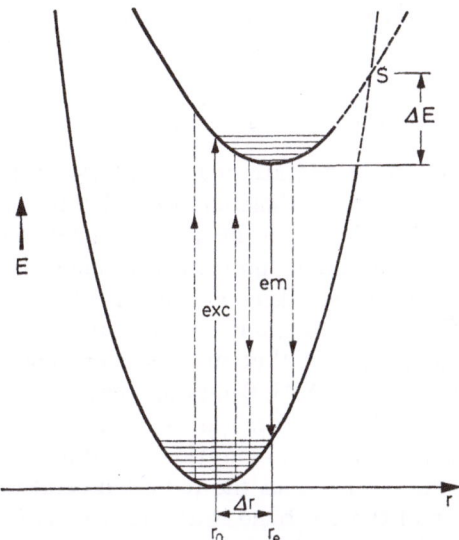

Fig. 6. Configuration coordinate diagram of a luminescent centre. Non-radiative return from the excited state to the ground state is possible via the crossover S. This requires an activation energy ΔE which can be supplied at higher temperatures. Exc: excitation, em: emission

Quenching of narrow-line emissions (as observed for many Ln^{3+} ions) has been explained by phonon emission to the lattice modes. *Moos* and co-workers (60) and others (61) have given many examples. Usually the nonradiative rate is described by Kiel's formula (62) for a single-frequency p-phonon process,

$$P_{nr} = A \cdot \varepsilon^{p} \left[\frac{\exp(\hbar\omega/kT)}{\exp(\hbar\omega/kT) - 1} \right]^{p},$$

where ω is the frequency of the emitted phonons, A is a constant and ε is a coupling constant. Note the completely different temperature dependence of temperature quenching of narrow-line and broad-band emission. The reader is referred to Refs. (60) and (63) for deviation of the formula and further discussion.

Very recently *Struck* and *Fonger* (64) have presented a unified model of temperature quenching of narrow-line and broad-band emissions. They use a quantum-mechanical single-configurational-coordinate model and were able to describe diverse types of quenchings with one and the same model. For a large Franck-Condon offset (*i.e.* Δr is large) the model gives approximately the single-activation-energy rate formula for the transition upwards to the crossower, but faster rates for the downward transitions. For small Franck-Condon offset with no relevant crossover the model gives a double-Poisson rate, which under certain conditions equals Kiel's multiphonon-emission rate. In passing we mention that *Lauer* and *Fong* (65) have also treated $f \rightarrow f$ and $d \rightarrow f$ relaxation of lanthanide ions starting from the same model. They present some interesting results on, among others, the phonons involved in the relaxation processes.

In this section we will discuss first the influence of the c.t. state of the Eu^{3+} ion on the temperature quenching of its luminescence, because it has been studied in detail. Secondly we will consider temperature quenching of some other lanthanides.

a) Temperature Quenching of the Eu^{3+} Emission

The first indication that the c.t. state of Eu^{3+} plays a role in the luminescence quenching process was the fact that there is a relation between the spectral position of the first c.t. band of Eu^{3+} and the quenching temperature and room-temperature quantum-efficiency of the luminescence under excitation into the c.t. band (66). A similar relation exists also for some other luminescent groups, *e.g.* the niobate octahedron $[NbO_6]^{7-}$ (67) and the uranate octahedron $[UO_6]^{6-}$ (68). *Bril* and co-workers (69) showed that at room temperature the luminescence quantum efficiency for Eu^{3+} in $YAl_3B_4O_{12}$ amounts to 35% for excitation into the c.t. band and to 100% for excitation into the narrow $4f$ levels. It is a simple task to show that in a simple configuration coordinate model the quenching temperature of the luminescence and the room-temperature quantum efficiency decrease, if the position of the c.t. band is at increasingly lower energy (67). This was the model to explain the results mentioned above.

The picture became more clear by the work of *Struck* and *Fonger* on temperature quenching of trivalent lanthanides in the oxysulfides (70, 71). In host lattices like Y_2O_2S and La_2O_2S the c.t. band of the Eu^{3+} ion is situated at about 30.000 cm^{-1} (70). This is lower than in the greater part of the oxides due to the lower electronegativity of sulfur. *Struck* and *Fonger* observed direct feeding of the excited $^5D(4f^6)$ levels of Eu^{3+} by the c.t. state, but also ^5D quenching via the c.t. state. They used a configuration coordinate diagram as given in Fig. 7. The important effect is that, although the c.t. state lies well above the emitting ^5D states in the absorption and excitation spectra, its Franck-Condon shifted minimum lies relatively low (somewhere near $^5D_3$). As a consequence crossovers from ^5D levels to the c.t. state are possible.

The direct contact between the c.t. state and the ^5D levels is shown by direct feeding of the ^5D levels by the c.t. state. If the Eu^{3+} ion in Y_2O_2S is excited into

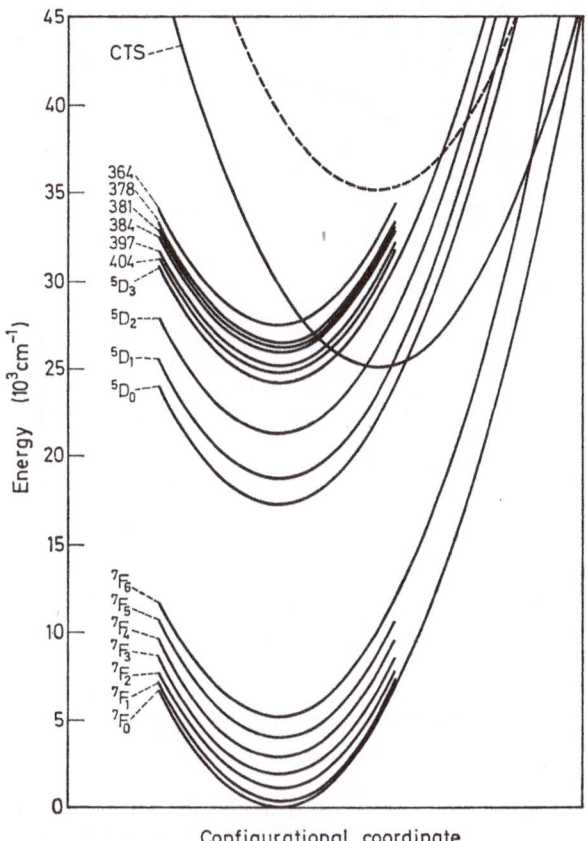

Fig. 7. Configuration coordinate model for the $4f^7$ and c.t. states (CTS) of Eu^{3+} in Y_2O_2S. The dotted curve shows qualitatively the higher position of the CTS in many oxidic hosts. The $4f$ states above $^5D_3$ are indexed by their absorption wavelengths (nm) from $^7F_0$. After Ref. (70)

the $^5L_7$ level, emission is observed from $^5D_3$, $^5D_2$, $^5D_1$ and $^5D_0$. The same emission spectrum is observed for excitation into the $^5D_3$ level. If excitation is into the c.t. state, *i.e.* at higher energies, emission occurs only from $^5D_2$, $^5D_1$ and $^5D_0$ in the same ratio as if the excitation had occurred into the $^5D_2$ level (70, 71). This means that c.t. excitation skips the $^5L_7$ and $^5D_3$ levels and feeds directly the $^5D_2$ level. In La_2O_2S—Eu, where the c.t. band is at still lower energy, the excited c.t. state feeds directly the $^5D_1$ level for about two-thirds and the $^5D_2$ level for about one-third.

In Fig. 8 the temperature dependence of the ^5D emissions in Y_2O_2S-Eu^{3+} for excitation into the c.t. state is given. Although the total emission intensity is practically temperature-independent the separate ^5D emissions quench sequentially in the order $^5D_3$, $^5D_2$, $^5D_1$, $^5D_0$ with increasing temperature. For La_2O_2S-Eu^{3+} the same sequence has been found, but the corresponding quenchings occur at lower temperatures. These quenchings are due to thermally promoted tran-

63

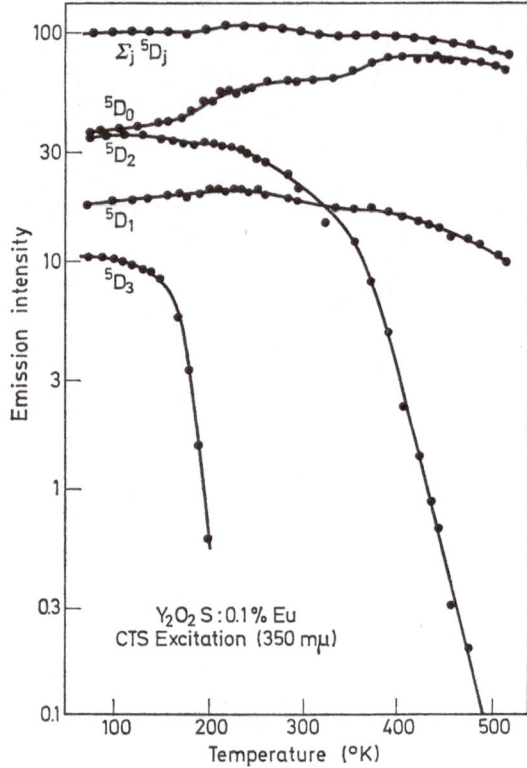

Fig. 8. Temperature dependence of the ^5D emissions of Eu^{3+} in Y_2O_2S-Eu(0.1%) for excitation into the c.t. state. The curves give the intensities of the four ^5D emission series expressed as percents of all emitted phonons. After Ref. (70)

sitions from the ^5D levels to the c.t. state followed by return crossovers to lower ^5D states. In the oxysulfides the c.t. state is low enough to allow such transitions. The crossover rates for c.t. state → ^5D levels are estimated to be 10^{11}–10^{12} sec^{-1}, so that the absence of luminescence from the c.t. state is understandable.

It will be clear that, if the c.t. state is at higher energy, these phenomena will no longer be observable.

In relation to this work on Eu^{3+} oxysulfides it is interesting to mention that it has been suggested that in these compounds the c.t. state dissociates into Eu^{2+} and a free hole (72). This suggestion is based upon the observation of energy storage and energy loss effects in the oxysulfides (73, 74). The energy storage effect can be induced by excitation into certain ^5D states. Although their life times are less than a millisecond, one can induce energy storage for days or weeks. The stored energy is released by thermal or infrared stimulation and reappears as normal Eu^{3+} emission. *Fonger* and *Struck* (74) have shown that the mechanism of this storage is as follows: The ^5D excitation reach the c.t. state through thermally activated crossover. The long storage time requests spatial separation of electrons and holes. Therefore the c.t. state dissociates into Eu^{2+} plus a free hole. These holes are stored in hole traps. After a storage time the hole traps can be emptied by thermal or infrared stimulation and enabled to return to the Eu^{2+}. The c.t. state is re-formed and c.t. state-to-^5D-feeding results in Eu^{3+} emission.

Special energy loss effects were observed at higher Eu^{3+} concentrations. It has been shown that these losses occur from the c.t. states. Again it is assumed that dissociation of the c.t. state is the first step to explain these concentration-dependent losses. At low Eu-concentration holes are freed and electrons remain immobilized at the originating Eu^{2+} centres; but at high Eu^{3+} concentrations the electrons can migrate through the europium ions to the trapped holes and recombine there nonradiatively. Storage is, therefore, a low-concentration effect, concentration-dependent loss a high-concentration phenomenon.

This model has been confirmed by the fact that La_2O_2S-Eu^{3+} shows p-type photoconductivity upon excitation into the $^5D_2$ and $^5D_3$ levels and the c.t. state (75).

After this discussion of the role of the c.t. states in the luminescence properties of Eu^{3+}-activated oxysulfides we now turn to Eu^{3+}-activated materials in general.

If the c.t. state is at relatively low energies we can expect phenomena like those described above for the oxysulfides. It has been shown that the temperature quenching of the $^5D_0$ emission of Eu^{3+} can occur via the c.t. state (76, 77). After $^5D_0$-c.t. crossover the c.t. state relaxes nonradiatively to the ground state as in the Mott-Seitz picture. It has been found that the thermal quenching temperature of Eu^{3+} emission under c.t. excitation increases if the c.t. state is situated at higher energies (78).

If the c.t. state lies at relatively high energy it is in fact impossible to quench the Eu^{3+} emission thermally if excitation is into the $4f$ levels (79). If excitation occurs into the c.t. state, the luminescence is quenched thermally, albeit at high temperatures. In Y_2O_3-Eu^{3+}, for example, with the c.t. state at about 50.000 cm^{-1}, the thermal quenching temperature of the Eu^{3+} emission is 800 K (78) for c.t. excitation and > 1000 K for $4f$ excitation (67). The structural requirements for a high thermal quenching temperature of Eu^{3+} emission under c.t. excitation have been given previously by us (79); if the lattice is rigid (small ions with high charge), Δr will be small and, therefore, the quenching temperature high. This has been verified experimentally. For example, the Eu^{3+} luminescence has a higher quenching temperature in $GdBO_3$ than in Gd_2O_3, in $LaBO_3$ than in $LaAlO_3$, in Lu-compounds than in the corresponding La-compounds. Many more examples have been reported.

These considerations are probably also of value in the study of Eu^{3+} luminescence in chelates. *Napier et al.* (80) have recently demonstrated the importance of the Eu^{3+} c.t. state for the absence of Eu^{3+} emission in tris (acetylacetonate) europium(III).

In a sence the well-known and extensively studied case of the $^2E \rightarrow {}^4A_2$ emission of Cr^{3+} in oxides (*e.g.* in ruby) can be treated in the same way. In comparison with Eu^{3+} the role of the emitting ^5D levels of the Eu^{3+} ion is played by the ^2E level of the Cr^{3+} ion and the role of the c.t. state of Eu^{3+} by the $^4T_2$ level of Cr^{3+}. Temperature quenching of the ^2E emission occurs via the $^4T_2$ level (81).

b) Temperature Quenching of Narrow-line Emission from other Lanthanide Ions

In the Eu^{3+} ion emission occurs between two $4f^6$ manifolds that are some 10.000 cm^{-1} apart. Multiphonon emission is, therefore, highly improbable. If the c.t. state is not accessible, the emission is practically "unquenchable" by thermal means. The same is true for the Tb^{3+} ion ($4f^8$). In fact the Tb^{3+} emission shows only temperature quenching, if the $4f^75d$ state is situated at low energy. *Struck* and *Fonger* (82) have shown that in La_2O_2S-Tb the $^5D_4$-Tb^{3+} emission under $^5D_4$ excitation is temperature-quenched via thermally promoted crossovers to Franck-Condon shifted states. For excitation into the $4f^75d$ state the situation is similar to that of Eu^{3+} (79).

For other lanthanides (except Gd^{3+}) multiphonon emission becomes more probable and the quenching mechanism has to be investigated from case to case [see *e.g.* (82)].

Finally it may be mentioned that other c.t. states can also play a role in the temperature quenching. The absence of Tb^{3+} emission in YVO_4 has been ascribed to the presence of a low-lying metal-metal c.t. state in which one of the Tb^{3+} electrons is transferred to the vanadate group (formally written as $Tb^{4+}+V^{4+}$). (30, 83). Assuming that this c.t. state has a large Franck-Condon shift it is easy to explain the absence of Tb^{3+} luminescence. Because one of the $4f$ electrons of Pr^{3+} is also easily excitable, similar phenomena are expected for Pr^{3+}. In fact Pr^{3+} in YVO_4 luminesces only very weakly.

In a similar way radiationless losses in other cases have been explained (virtual recharge, see above).

c) Temperature Quenching of Broad-band Emission from Lanthanide Ions

A number of lanthanide ions show Stokes-shifted broad-band emission. These are usually $5d \rightarrow 4f$ transitions. Examples are $Ce^{3+}(4f^1)$ and $Eu^{2+}(4f^7)$. Their temperature quenching can be described with the Mott-Seitz picture (see Fig. 6). The properties of the $4f^{n-1}5d$ state determine the value of the quenching temperature of the luminescence; the value of Δr should be small and the position of the $4f^{n-1}5d$ at high energy. This explains, among others, why Ce^{3+} and Eu^{2+} luminesce efficiently in strongly bound oxidic host lattices (*e.g.* silicates, phosphates) (79). Data on ions with broad-band $5d \rightarrow 4f$ emission can be used to predict the temperature quenching of ions with narrow-line emission under $4f \rightarrow 5d$ excitation: the quenching temperature of this emission is determined by the characteristics of the $4f^{n-1}5d$ state. As an example we mention that under $4f \rightarrow 5d$ excitation the Tb^{3+} ion shows high quenching temperatures in the same lattices where the Ce^{3+} ions shows high quenching temperatures of its emission.

d) Emission from Tetravalent Lanthanides?

In his study on Ln^{4+} ions in oxides *Hoefdraad* (5) did not observe Ln^{4+} luminescence, not even at 4 K. The lowest excited state in these ions is a c.t. state (see Section 2). It is a little surprising that no clear example has been found of luminescence from a c.t. state in which a $4f$ state is involved, whereas luminescence from c.t. states in which a d state is involved is quite common (tungstates, niobates, vanadates). Note that luminescence from a c.t. state involving a $5f$ state has also been observed, viz. on hexavalent uranium in oxides (see Section 6).

It has been proposed (5) that in the case of the Ln^{4+} ions an excited state involving a metal d state is responsible for the absence of luminescence. It is then necessary that this state has a large Franck-Condon shift and is situated at energies not very much higher than those of the $4f$-c.t. state. Whatever the solution of this problem may be, it is clear that either c.t. or $4f^{n-1}5d$ states play an important role in the quenching process of the luminescence. The conclusion of all the material presented in this section is that this is true for all types of lanthanide luminescence.

5. Influence of Charge-transfer and $4f^{n-1} 5d$ States on Energy Transfer Probabilities

Energy transfer is the phenomenon that a certain ion or a group of ions (more generally a centre) that has been raised into an excited state transfers its excitation energy (or part of this energy) to another centre which can emit the energy radiatively. Energy transfer is often of great importance in luminescent materials, especially those containing lanthanides, for the following reason. A first requirement for a luminescent material is that it emits radiation. It is trivial that this leads to the requirement that exciting radiation must be absorbed. The $4f \rightarrow 4f$ transitions of the lanthanides are not very suitable for absorption of radiation, because they are strongly forbidden. Excitation may occur efficiently in either the c.t. state or the $4f^{n-1} 5d$ state. An example of the first case is Gd_2O_3-Eu^{3+} (84). Short-wave UV radiation is strongly absorbed by the Eu^{3+} ion raising this ion into its c.t. state. The ion then relaxes to the $^5D_0$ level from which luminescence occurs. An example of the second case is $YTaO_4$-Tb^{3+} which can be excited efficiently by UV radiation into the $4f^7 5d$ state (38).

It is also possible, however, to excite the lanthanide ions indirectly. This is done by building into the lattice another ion or group of ions that absorbs strongly the exciting radiation and, subsequently, transfer this energy to the lanthanide ions.

We give two examples. The Tb^{3+} ion in $YAl_3B_4O_{12}$ cannot be excited by 254 nm radiation (from a low-pressure mercury lamp), because Tb^{3+} in this lattice does not absorb this radiation. Ce^{3+} in $YAl_3B_4O_{12}$, however, does absorb this radiation. Energy transfer from Ce^{3+} to Tb^{3+} occurs so that excitation into the Ce^{3+} ion is followed by Tb^{3+} emission (85).

Our second example concerns Eu^{3+} in YPO_4. In this lattice Eu^{3+} absorbs 254 nm only to a minor amount. If part of the phosphorus is replaced by vanadium $[(Y,Eu)P_{1-x}V_xO_4]$ the radiation is absorbed by the vanadate group and transferred to the Eu^{3+} ion.

What is the mechanism by which this energy transfer occurs (86)? Figure 9 shows that should happen. We start with the system S (excited)$+$ A (ground state) and end up with the system S (ground state) $+$A (excited). This is only possible, if one of the levels of A lies at the same height as the luminescent level of S (resonance condition). Further we need an interaction between S and A.

Transfer can be brought about in the first place by the Coulomb interaction between all charged particles of S and A. If S and A are so far apart that their charge clouds do not overlap, this form of energy transfer is the only one possible. If these do overlap, however, another process is possible by exchange interaction between the electrons of S and A. In this process electrons are exchanged between S and A, in the former the electrons remain with their respective ions.

This may be summarized as follows. For transfer by Coulomb interaction we write

$$P_{SA} = g_{SA} \cdot E_{SA}$$

and for transfer by exchange interaction

$$P_{SA} = f_{SA} \cdot E_{SA}.$$

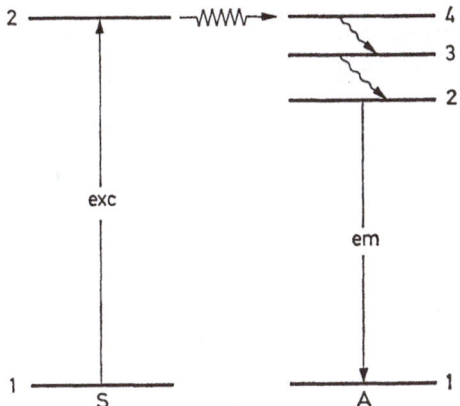

Fig. 9. Energy transfer from S to A. Transfer occurs to level 4, followed by radiationless decay to level 2 from which emission occurs

Here P_{SA} is the transfer probability. E_{SA} represents the resonance condition (in practise the spectral overlap of the emission of S and the relevant absorption of A) and occurs in both formulas. The quantity g_{SA} comprises the optical strengths of the relevant transitions and a distance-dependence of the type r_{SA}^{-n} ($n=6,8$, etc.). The quantity f_{SA}, however, is proportional to the wave function overlap of S and A and comprises an exponential distance-dependence.

It will be obvious that the spectral position of the c.t. or $4f_n^{-1}5d$ state of the lanthanide ions may influence P_{SA} strongly. On the other hand the presence of such a state does not lead to unexpected new phenomena. This will be illustrated on the example of the Eu^{3+} ion in $YNbO_4$ and $YTaO_4$ (*38*). In both host lattices we can excite the luminescent host lattice groups NbO_4^{3-} and TaO_4^{3-}. It can be shown that the host lattice excitation energy does not migrate through the lattice. Transfer from the niobate group to Eu^{3+} occurs only over distances shorter than about 4 Å; transfer from the tantalate group to Eu^{3+} occurs over distances up to about 10 Å. This difference is due to the fact that the niobate emission with a maximum at about 400 nm overlaps only with Eu^{3+} $4f \rightarrow 4f$ absorption bands. Due to their low oscillator strength P_{SA} will be small (probably exchange interaction). The tantalate emission, however, has its maximum at 335 nm and overlaps the Eu^{3+} c.t. absorption band (maximum at 245 nm, but extending down to 300 nm). The $TaO_4 \rightarrow Eu^{3+}$ transfer occurs by dipole-dipole interaction and has a higher probability than the $NbO_4 \rightarrow Eu^{3+}$ transfer.

Finally we want to illustrate the influence of c.t. states on transfer probabilities using the example of energy transfer from the VO_4^{3-} group to trivalent lanthanides (*87, 88*). These have been studied in YVO_4, $CaSO_4$ and $PbSO_4$. In the sulfates the V^{5+} and Ln^{3+} ions from associates. Table 8 surveys the results qualitatively. Note the low transfer probabilities in $PbSO_4$ and the different transfer probabilities for $VO_4^{3-} \rightarrow Tb^{3+}$ and Pr^{3+} in YVO_4 and $CaSO_4$.

The absence of Tb^{3+} and Pr^{3+} luminescence in YVO_4 has been ascribed to the presence of low-lying c.t. states of the type $V^{4+}-Ln^{4+}$ from which the system is

Table 8. Energy transfer probabilities for associates $VO_4^{3-}Ln^{3+}$ studied in YVO_4, $CaSO_4$ and $PbSO_4$ ($-$: no transfer, $+$: efficient transfer, 0: transfer probability low to medium)

	YVO_4	$CaSO_4$	$PbSO_4$
$VO_4^{3-} \rightarrow Pr^{3+}$	$-$	$+$	0
$VO_4^{3-} \rightarrow Sm^{3+}$	$+$	$+$	0
$VO_4^{3-} \rightarrow Eu^{3+}$	$+$	$+$	0
$VO_4^{3-} \rightarrow Tb^{3+}$	$-$	$+$	0
$VO_4^{3-} \rightarrow Dy^{3+}$	$+$	$+$	0

assumed to return nonradiatively to the ground state (see Section 4b). It is reasonable that such an effect would occur for terbium and praseodymium with relatively low fourth ionization potentials. In the sulfates this c.t. state should be situated at much higher energy, because the $Tb^{3+}(Pr^{3+})$ ion (on a Me^{2+} ion site) is effectively positive, whereas the V^{5+} ion (on a S^{6+} site) is effectively negative. Electron transfer in such a situation is expected to take more energy than in YVO_4 without effective charges.

The inefficiency of energy transfer from the VO_4^{3-} group to trivalent lanthanides in $PbSO_4$ may be due to the fact that the lowest excited state of the VO_4^{3-} group in $PbSO_4$ has strong Pb^{3+}—V^{4+} c.t. character. Its life time is about thirty times longer than for the lowest excited state of VO_4^{3-} in $CaSO_4$. It may be expected that such a Pb^{3+}—V^{4+} c.t. state has considerably less wavefunction overlap with the lanthanide $4f$-orbitals than the normal c.t. state of the VO_4^{3-} group. As a consequence P_{SA} in $PbSO_4$ would be much lower than in $CaSO_4$. These transfer mechanisms have certainly not been elucidated completely. The latter example, however, indicates clearly that the influence of c.t. states on energy transfer probabilities may be drastic.

6. The Luminescence of Hexavalent Uranium in Oxides

Of all the actinides the luminescence of the U^{6+} ion has been investigated most thoroughly. The spectra of the uranyl ion (UO_2^{2+}) have received much attention [for a review see Ref. (89)] and will not be discussed here. The uranyl complexes are highly anisotropic, since two oxygen ions are bound at short distances, whereas others are much more weakly bound at larger distances. Sometimes uranium luminescence from solids is ascribed to the uranyl ion without further discussion. This, however, is incorrect because there is strong evidence that this emission can also occur from octahedral UO_6^{6-} and tetrahedral UO_4^{2-} groups (90). These groups are isotropic. We will consider some recent results on the UO_6^{6-} octahedron because this example is easy to discuss, shows some clear differences with the uranyl ion and also illustrates the main theme of this paper.

How can we be sure that the $U^{6+}(O^{2-})_n$ complex in a mixed metal oxide is present as the UO_6^{6-} octahedron? This can be done by studying solid solution series between tungstates (tellurates, etc.) and uranates which are isomorphous and whose crystal structure is known. Illustrative examples are solid solution series with ordered perovskite structure $A_2BW_{1-x}U_xO_6$ and $A_2BTe_{1-x}U_xO_6$ (91). Here A and B are alkaline-earth ions. The hexavalent ions occupy octahedral positions as can be shown by infrared and Raman analysis (92, 93). Usually no accurate determinations of the crystallographic anion parameters are available, because this can only be done by neutron diffraction [see however Ref. (94)]. Vibrational spectroscopy is then a simple tool to determine the site symmetry of the uranate complex in the lattice, if these groups do not have oxygen ions in common. In the perovskite structure this requirement is fulfilled.

Fortunately these perovskites show an efficient luminescence upon excitation with UV radiation (91, 95, 96). The emission of Ba_2MgWO_6-U is given in Fig. 10

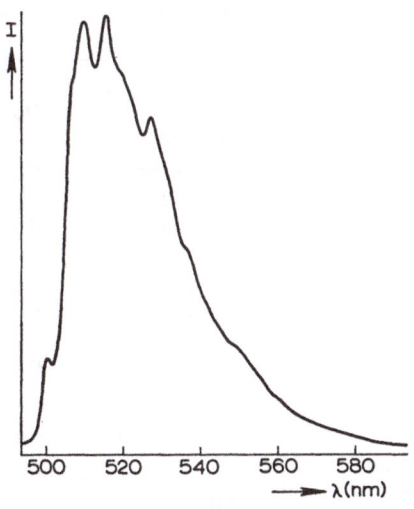

Fig. 10. Spectral energy distribution of Ba_2MgWO_6-U (0.3%)

as an example. This emission shows pronounced vibrational structure which differs considerably from the fine structure in the case of the uranyl complex. In the latter case the most intense features correspond to a two-phonon process and the vibrational pattern is usually observed to repeat several times (97). In the case of the UO_6^{6-} octahedron the one-phonon part of the side band is well defined and two-phonon lines are weak. This indicates that in the uranyl complex the electron-phonon coupling is stronger than in the UO_6^{6-} complex.

From life-time measurements it can be concluded that the emission transition is strongly forbidden as an electric-dipole transition. In the ordered perovskites the radiant decay time amounts to about 300 μsec (98). In Table 9 we have presented data on decay times of UO_6^{6-} luminescence which show that the decay time is about 200—300 μsec if the U^{6+} ion occupies a centre of inversion in the crystal lattice, but that it becomes considerably shorter if there is no inversion symmetry at the site of the U^{6+} ion. This seems to indicate that the transition involved is parity-forbidden. It is interesting in this connection that *Runciman* and co-workers (102) have reported that for a uranate centre in CaF_2 (possibly UO_6^{6-}) the zero-phonon emission line is of a magnetic-dipole origin. Furthermore it has been found (98) that the life time of the UO_6^{6-} emitting level decreases with increasing temperature, whereas the luminescence intensity remains constant in the temperature region involved. Such a behaviour is to be expected for a parity-forbidden transition, since the probability for vibronically induced transitions will increase with increasing temperature. As a consequence the vibrational fine structure of the emission fades away at higher temperatures.

Table 9. Decay time of the UO_6^{6-} luminescence at 77 K as a function of the crystallographic site symmetry

Composition	Decay time[a] (μsec)	Crystal structure ordered	Site symmetry	Ref.
Ba_2MgWO_6-U	250	ordered perovskite	O_h	(98)
Ba_2ZnWO_6-U	240	ordered perovskite	O_h	(98)
Ba_2SrWO_6-U	200	ordered perovskite	slight deviation from O_h	(98)
$SrLaNaWO_6$-U	155	ordered perovskite	slight deviation from O_h	(98)
$Y_3Li_3Te_2O_{12}$-U	240	garnet	S_6	(99)
Ba_2TeO_5-U	42	Ba_2WO_5	no inversion symmetry	(100)
$BaLi_{0.4}W_{0.6}O_3$-U	80	own type with two W-sites	C_{3v}	(101)
	10		C_{3v}	

[a]) Values at liquid nitrogen temperature. Values at liquid helium temperature are even longer.

There is convincing evidence that the transition involved is a c.t. transition between the oxygen $2p$-orbitals and the empty uranium $5f$-orbital ($89, 103$). This means that the energy level scheme mentioned on page 47 in the discussion of octahedrally coordinated tetravalent lanthanide ions may also be applicable in the case of the UO_6^{6-} group. The highest occupied orbital in that scheme has t_{1u} symmetry, whereas the lowest unoccupied orbital is the $5f$ orbital with "ungerade" symmetry, so that in this scheme transitions between the two levels are in fact parity-forbidden. A description along these lines explains the long decay time of the emission and the influence of inversion symmetry. The occurrence of a zero-phonon line with vibronic side bands indicates that the Franck-Condon shift of the excited state relative to the ground state is rather small. This may be connected with the weakly-antibonding character of the $5f$ orbitals. Note that emission from c.t. states involving d-states is always broad-band emission without any structure (i.e. large Franck-Condon shift).

In the case of the Ln^{4+} ions the spectral position of the transitions was found to be independent of the host lattice. It is interesting to investigate this dependence in the case of the UO_6^{6-} group, because the $5f$ orbitals are known to behave in between the $3d$ electrons of the first transition series and the $4f$ electrons of the lanthanides ($13, 103, 104$). Contrary to the case of the Ln^{4+} ions the transition as observed from the emission spectra of the uranate group is host-lattice dependent in a rather drastic way: representative border-cases are the 510 nm (green) emission from Ba_2MgWO_6-U (97) and the 590 nm (orange) emission of Ba_2TeO_5-U (96). For tungstate luminescence involving a $5d$ state the magnitude of the variation is similar (105).

In a rough model one might expect that, if the UO_6^{6-} group is surrounded by more electronegative cations (i.e. if the bonding in UO_6^{6-} is relatively ionic), the emission is more towards the yellow. Examples in Table 10 show that there is some experimental evidence for such a model. Independently of the question whether this model is correct or not, it is clear that the position of the c.t. transi-

Table 10. Position of the maximum of the UO_6^{6-} emission at room temperature in several host lattices as a function of the next-nearest cation neighbours

Composition	Crystal structure	Maximum (nm)	Cation neighbours	Ref.
Ba_2MgWO_6-U	ordered perovskite	510	Ba^{2+}, Mg^{2+}	(97)
Li_4WO_5	"rock salt like"	520	Li^+, W^{6+}	(106)
Ba_2ZnWO_6-U	ordered perovskite	535[a]	Ba^{2+}, Zn^{2+}	(98)
Y_6WO_{12}-U	"fluorite-like"	540[a]	Y^{3+}	(90)
$Y_3Li_3Te_2O_{12}$-U	garnet	560	Li^+, Y^{3+}	(99)
Ba_2WO_5-U	own type	560	Ba^{2+}, W^{6+}	(107)
$BaLi_{0.4}W_{0.6}O_3$-U	own type	560	Ba^{2+}, Li^+, W^{6+}	(101)
$MgWO_4$-U	wolframite	590/610	Mg^{2+}, W^{6+}	(105)

[a]) Values at liquid nitrogen temperature.

tion involving the $5f$ orbital of U^{6+} is sensitive to the surroundings in the lattice. This sensitivity is more pronounced than in the case of the uranyl ion.

The luminescence characteristics (fine structure, decay time) depend, therefore, strongly on the fact whether the d or the f state of the central metal ion is involved in the c.t. state, although the dependence of the emission maximum on the host lattice is similar. Such drastic differences are also observed for magnetic properties of ordered perovskites: whereas Ba_2NiWO_6 is antiferromagnetic (108), Ba_2NiUO_6 is ferromagnetic (109). The WO_6 and UO_6 group play the role of the intermediary in the superexchange interaction between the Ni^{2+} ions. The d-functions with even parity lead to antiferromagnetic interactions, the f-functions with uneven parity to ferromagnetic interactions between the Ni^{2+} ions.

After the emission of the UO_6^{6-} group we finally turn to the absorption spectra. In the ordered perovskites the absorption transition corresponding to the emission discussed above can be found from excitation spectra of the luminescence (98) and from diffuse reflection spectra of the compounds A_2BUO_6 (93). As is to be expected its absorption strength is relatively low, but no quantitative data are available at the moment. The corresponding emission shows only a small Stokes shift (a few kK) as is to be expected in view of the vibronic structure observed. The UO_6^{6-} emission shows concentration quenching at low uranium concentrations [1 atomic percent or less (98, 99)] which must be related to the small Stokes shift [large spectral overlap, high transfer probability (86)].

In the near ultraviolet, however, a more intense absorption band is present (98, 99, 107). Obvious assignments for this band are

a) c.t. from filled "gerade" molecular orbitals of the oxygen ions to the $5f$ orbitals of the uranium ion.

b) c.t. involving the $6d$ level of the uranium ion.

There are arguments in favour both assignments and it seems impossible at the moment to come to a definite conclusion. Although this band has sometimes some broad shoulders, we never observed any indication of vibronic fine structure, not even at liquid helium temperature (98, 99). This is in favour of the $6d$ level but on the other hand a transition originating from lower molecular orbitals (to $5f$) may lack fine structure due to the higher bonding character of the initial m.o. which results in a larger difference between the equilibrium internuclear distances (Franck-Condon shift). In addition there is a correlation between the positions of the W^{6+} c.t. transitions and of the strong U^{6+} band in a series of different crystal lattices (see Table 11). The difference between these positions amounts to 5—6 kK. Such a relation is to be expected if the $6d$ orbital is involved. In view of the similar spectral variation of the U^{6+} emission (involving a $5f$ orbital) and the W^{6+} emission (involving a $5d$ orbital) this argument is not convincing too. The fact that U^{6+} in the garnet $Y_3Li_3Te_2O_{12}$ (99) has its emission at relatively long wavelength (580 nm), but the intensive excitation band in the ultraviolet at relatively short wavelength (325 nm, see Table 11) may be an indication that the $6d$ level is involved in the intense excitation band. The reverse behaviour of emission and excitation is otherwise difficult to explain.

Table 11. Position of the first strong excitation bands of the luminescence of the WO_6^{6-} and the UO_6^{6-} octahedra in several host lattices. Values in kK

Host lattice (reference)	Position excitation bands		Difference
	WO_6^{6-}	UO_6^{6-}	
$Y_3Li_3Te_2O_{12}$ (99)	37	31	6
Ba_2MgWO_6 (98)	32	27	5
$BaLi_{0.4}W_{0.6}O_3$ (101)	32	~26	~6
Ba_2WO_5 (107)	31	25.5	5.5
$BaNa_{0.4}W_{0.6}O_3$ (110)	~29	24	~5

It is further noteworthy that the quenching temperature of the uranate emission for excitation into the strong excitation band decreases from top to bottom in Table 11 depending on the position of this excitation band. The same is true for the tungstate luminescence in the compounds cited. This seems to indicate that the temperature quenching of the uranate emission that originates from the c.t. state involving the $5f$ orbital occurs via the c.t. state involving the $6d$ orbital. Further investigations are necessary, but the influence of c.t. states upon the luminescence properties of the UO_6^{6-} group is obvious.

7. Conclusions

If we survey the reviewed data it is clear that the luminescence properties of the lanthanides (and also the actinides) are often strongly influenced by either charge-transfer or $f^{n-1}d$ states if these states are situated at not too low energy. The work by *Struck* and *Fonger* (70—72) is a good example of a quantitative approach to problems of this kind. Their case of the trivalent lanthanides in the oxysulfides is by no means exceptional as shown in this review. We expect that in the future a number of phenomena related to luminescence will be explained along lines which have been indicated above. A serious problem will always be the position of the c. t. or $f^{n-1}d$ state relative to the ground state in the potential energy vs configuration coordinate plot. The Franck-Condon shift between these two states will in general result in an extra degree of freedom in the interpretation of the experimental results.

8. References

1. *Jørgensen, C. K.:* Mol. Phys. *5*, 271 (1962).
2. *Nugent, L. J., Baybarz, R. D., Burnett, J. L., Ryan, J. L.:* J. Phys. Chem. *77*, 1528 (1973).
3. *Jørgensen, C. K., Rittershaus, E.:* Matematisk-fysiske Meddelelser Det Kongelige Danske Videnskabernes Selskab *35*, no. 15 1967.
4. *van Vugt, N., Wigmans, T., Blasse, G.:* J. Inorg. Nucl. Chem. *35*, 2602 (1973).
5. *Hoefdraad, H. E.;* J. Inorg. Nucl. Chem., *37*, 1917 (1975); thesis, Utrecht, 1975.
6. *Schmidtke, H. H.:* J. Chem. Phys. *45*, 3920 (1966).
7. *Hoefdraad, H. E., Blasse, G.:* unpublished.
8. *Jørgensen, C. K.:* Progr. Inorg. Chem. *12*, 101 (1970).
9. *McClure, D. S., Kiss, Z. J.:* J. Chem. Phys. *39*, 3251 (1963).
10. *Blasse, G., Wanmaker, W. L., ter Vrugt, J. W.:* J. Electrochem. Soc. *115*, 673 (1968) and references cited there in.
11. *Wood, D. L., Kaiser, W.:* Phys. Rev. *126*, 2079 (1962). — *Kiss, Z. J., Weakliem, H. A.:* Phys. Rev. Letters *15*, 457 (1965). — *Mahbub'ul Alam, A. S. M., Di Bartolo, B.:* J. Chem. Phys. *47*, 3790 (1967).
12. *Palilla, F. C., O'Reilly, B. E., Abbruscato, V. J.:* J. Electrochem. Soc. *117*, 87 (1970). — *Witzke, H., McClure, D. S., Mitchell, B.:* Luminescence of crystals, molecules and solution (Proc. Int. Conf. Leningrad, 1972), ed. *F. Williams.* New York: Plenum Press 1973.
13. *Jørgensen, C. K.:* Modern aspects of ligand field theory, North Holland, Amsterdam, 1971; Progr. Inorg. Chem. *4*, 73 (1962).
14. *Kaplyanski, A. A., Feofilov, P. P.:* Opt. Spectry. (USSR) (English Transl.) *13*, 129 (1962); (Opt. i Spektroskopiay *13*, 235 (1962)).
15. *Kirs, J., Niilks, A.:* Tr. Inst. Fiz. i Astron. Akad. Nauk Est. SSR *18*, 36 (1962).
16. *Reisfeld, R., Grabner, A.:* J. Opt. Soc. Am. *54*, 331 (1964).
17. *Blasse, G., Wanmaker, W. L., ter Vrugt, J. W., Bril, A.:* Philips Res. Rep. *23*, 189 (1968).
18. *Loh, E.:* Phys. Rev. *147*, 332 (1966).
19. *Barnes, J. C., Pincott, H.:* J. Chem. Soc. *1966*, 842.
20. *Day, P., Diggle, P. J., Griffiths, G. A.:* J. Chem. Soc. Dalton Transact. *1974*, 1446.
21. *Blasse, G.:* J. Solid State Chem. *4*, 52 (1972).
22. *Hoefdraad, H. E.;* J. Solid State Chem., *15*, 175 (1975).
23. *Blasse, G., Bril, A.:* J. Chem. Phys. *47*, 5139 (1967).
24. *Loh, E.:* Phys. Rev. *154*, 270 (1967).
25. *Hoshina, T., Kuboniwa, S.:* J. Phys. Soc. Japan *31*, 828 (1971). — *Nakazawa, E., Shionoya, S.:* J. Phys. Soc. Japan *36*, 504 (1974).
26. *Ryan, F. M., Lehman, W., Feldman, D. W., Murphy, J.:* J. Electrochem. Soc. *121*, 1475 (1974).
27. *McClure, D. S., et al.:* Ref. 12.
28. *Walker, W., Butler, K. H.:* J. Electrochem. Soc. *116*, 1245 (1969).
29. *van Loo, W.:* Phys. Stat. Sol. (a) *27*, 565 (1975); *28*, 227 (1975).
30. *Blasse, G., Bril, A.:* Philips Res. Rep. *22*, 481 (1967).
31. *Haas, Y., Stein, G., Fonkiewicz, M.:* J. Phys. Chem. *74*, 2558 (1970).
32. *Ofelt, G. S.:* J. Chem. Phys. *38*, 2171 (1963).
33. *Blasse, G., Bril, A.:* J. Chem. Phys. *50*, 2974 (1969).
34. *Reisfeld, R., Lieblich, N.:* J. Phys. Chem. Solids *34*, 1467 (1973).
35. *Reisfeld, R., Eckstein, Y.:* Solid State Commun. *13*, 265 (1973).
36. *Hoshino, T.:* Jap. J. Appl. Phys. *6*, 1203 (1967).
37. *Avella, F. J.:* J. Electrochem. Soc. *113*, 1225 (1966).
38. *Blasse, G., Bril, A.:* J. Luminescence *3*, 109 (1970).
39. *Jørgensen, C. K., Judd, B. R.:* Mol. Phys. *8*, 281 (1964).
40. *Blasse, G., Bril, A., Nieuwpoort, W. C.:* J. Phys. Chem. Solids *27*, 1587 (1960).
41. *Judd, B. R.:* J. Chem. Phys. *44*, 839 (1966).
42. *Blasse, G., Bril, A.:* Solid State Commun. *4*, 373 (1966).
43. *Blasse, G., Bril, A.:* Philips Res. Rep. *21*, 368 (1966). — *Nieuwpoort, W. C., Blasse, G.:* Solid State Commun. *4*, 227 (1966).
44. *Mason, S. F., Peacock, R. D., Stewart, B.:* Chem. Phys. Letters *29*, 149 (1974).

45. *Piper, W. W., DeLuca, J. A., Ham, F. S.:* J. Luminescence *8*, 344 (1974).
46. *Sommerdijk, J. L., Bril, A., de Jager, A. W.:* J. Luminescence *8*, 341 (1974).
47. *Sommerdijk, J. L., Bril, A., de Jager, A. W.:* J. Luminescence *9*, 288 (1974).
48. *Weber, M. J.:* Solid State Commun. *12*, 741 (1973).
49. *Reut, E. G., Ryskin, A. I.:* Phys. Status Solidi (a) *17*, 47 (1973); Opt. i. Spectroskopiya *35*, 862 (1973).
50. *Hoefdraad, H. E., Blasse, G.:* Phys. Status Solidi (a) *29*, K95 (1975).
51. *Hewes, R. A., Hoffmann, M. V.:* J. Luminescence *3*, 261 (1971).
52. *Hoffmann, M. V.:* J. Electrochem. Soc. *118*, 933 (1971); *119*, 905 (1972).
53. *Blasse, G.:* Phys. Status Solidi (b) *55*, K 131 (1973).
54. *Verstegen, J. M. P. J., Sommerdijk, J. L.:* J. Luminescence *9*, 297 (1974).
55. *Tangry, B., Pezat, M., Fontenit, C., Fouassier, C.:* Compt. Rend. Acad. Sci. Paris *277*, 25 (1973).
56. *Sommerdijk, J. L., Verstegen, J. M. P. J., Bril, A.:* J. Luminescence *8*, 502 (1974).
57. *Verstegen, J. M. P. J., Sommerdijk, J. L., Bril, A.:* J. Luminescence *9*, 420 (1974).
58. *Klick, C. C., Schulman, J. H.:* Solid State Phys. *5*, 97 (1957).
59. *Curie, D.:* Luminescence in crystals. New York: Wiley 1963.
60. *Moos, H. W.:* J. Luminescence *1*, *2*, 106 (1970).
61. *Weber, M. J.:* Phys. Rev. *171*, 283 (1968). — *Chamberlain, J. R., Paxman, D. H., Page, J. L.:* Proc. Phys. Soc. *89*, 143 (1966).
62. *Kiel, A.:* Proc. Third Internat. Conf. Quantum Electronics, Paris, 1963 (*P. Grivet* and *N. Bloembergen,* eds.), p. 765. New York: Columbia University Press 1964.
63. *Riseberg, L. A., Weber, M. J.:* Progress in Optics, in press.
64. *Struck, C. W., Fonger, W. H.:* J. Luminescence *10*, 1 (1975).
65. *Lauer, H. V., Fong, F. K.:* J. Chem. Phys. *60*, 274 (1974).
66. *Blasse, G.:* J. Chem. Phys. *45*, 2356 (1963).
67. *Blasse, G.:* J. Chem. Phys. *48*, 3108 (1968).
68. *Blasse, G.:* J. Electrochem. Soc. *115*, 738 (1968).
69. *Bril, A., Blasse, G., Bertens, J. A. A.* : J. Electrochem. Soc. *115*, 395 (1968).
70. *Struck, C. W., Fonger, W. H.:* J. Luminescence *1*, *2*, 456 (1970).
71. *Struck, C. W., Fonger, W. H.:* J. Chem. Phys. *52*, 6364 (1970).
72. *Struck, C. W., Fonger, W. H.:* Phys. Rev. B *4*, 22 (1971).
73. *Forest, H., Cocco, A., Hersh, H.:* J. Luminescence *3*, 25 (1970).
74. *Fonger, W. H., Struck, C. W.:* J. Electrochem. Soc. *118*, 273 (1971).
75. *Dobrov, W. I., Buchanan, R. A.:* Appl. Phys. Letters *21*, 201 (1972).
76. *Blasse, G., Bril, A., de Poorter, J. A.:* J. Chem. Phys. *53*, 4450 (1970).
77. *Delsart, Ch.:* J. Physique *34*, 711 (1973).
78. *Blasse, G., de Vries, J.:* J. Electrochem. Soc. *114*, 875 (1967).
79. *Blasse, G.:* J. Chem. Phys. *51*, 3529 (1969).
80. *Napier, G. D. R., Neilson, J. D., Shepherd, T. M.:* Chem. Phys. Letters *31*, 328 (1975).
81. See *Castrelli, F., Forster, L. S.:* Phys. Rev. B *11*, 920 (1975). — *Fonger, W. H., Struck, C. W.:* Phys. Rev. B *11*, 3251 (1975).
82. *Struck, C. W., Fonger, W. H.:* J. Appl. Phys. *42*, 4515 (1971).
83. *De Losh, R. G., Tien, T. Y., Gibbons, E. F., Zacmanides, P. J., Stadler, H. L.:* J. Chem. Phys. *53*, 681 (1970).
84. *Bril, A., Wanmaker, W. L.:* J. Electrochem. Soc. *111*, 1363 (1964).
85. *Blasse, G., Bril, A.:* J. Chem. Phys. *47*, 1920 (1967).
86. For a full acount see *Dexter, D. L.:* J. Chem. Phys. *21*, 836 (1953).
87. *Draai, W. T., Blasse, G.:* Phys. Status Solidi (a) *21*, 569 (1974).
88. *Blasse, G., van den Heuvel, G. P. M.:* J. Luminescence, *10* (1975).
89. *Burrows, H. D., Kemp, T. J.:* Chem. Soc. Rev. *1974*, 139.
90. *Blasse, G.:* J. Electrochem. Soc. *115*, 738 (1968).
91. *de Hair, J. Th. W., Blasse, G.:* J. Luminescence *8*, 97 (1973).
92. *Blasse, G., Corsmit, A. F.:* J. Solid State Chem. *6*, 513 (1973).
93. *Kemmler-Sack, S., Seemann, I.:* Z. Anorg. Allgem. Chem. *411*, 61 (1975).
94. *Rietveld, H. M.:* Acta Cryst. *20*, 508 (1966). — *Loopstra, B. O., Rietveld, H. M.:* Acta Cryst. B *25*, 787 (1969).

95. *Runciman, W. A.:* Brit. J. Appl. Phys. Suppl. *4*, S 78 (1955).
96. *Natansohn, S.:* J. Electrochem. Soc. *120*, 660 (1973).
97. *Dieke, G. H., Duncan, A. B. F.:* Spectroscopic properties of uranium compounds, New York: McGraw-Hill 1949.
98. *de Hair, J. Th. W., Blasse, G.:* Chem. Phys. Letters *36*, 111 (1975).
99. *Alberda, R. H., Blasse, G.:* J. Luminescence, in press.
100. *van den Heuvel, G. P. M., Blasse, G.:* unpublished results.
101. *Blasse, G.:* J. Solid State Chem. *14*, 366 (1975).
102. *Manson, N. B., Shah, G. A., Runciman, W. A.:* Solid State Commun. *16*, 645 (1975).
103. *Jørgensen, C. K.:* Proc. Symposium Tihany 1964, Budapest, Akadémia Kiado, 1965, p. 11. — *Jørgensen, C. K., Reisfeld, R.:* Chem. Phys. Letters *35*, 441 (1975).
104. *Boring, M., Wood, J. H., Moskowitz, J. W.:* J. Chem. Phys. *61*, 3800 (1974).
105. *Kröger, F. A.:* Some aspects of the luminescence of solids. Amsterdam: Elsevier Publ. Comp. 1948.
106. *Leonov, Yu. S.:* Opt. Spectry. (USSR) (English Transl.) *9*, 145 (1960); *10*, 357 (1961).
107. *Blasse, G., van den Heuvel, G. P. M.:* J. Luminescence *8*, 406 (1974).
108. *Blasse, G.:* Proc. Int. Conf. Magnetism, Nottingham, 1964, p. 350, Inst. Phys., London (1965).
109. *Scharf, W., Weitzel, H.:* Solid State Commun. *15*, 1831 (1974).
110. *Bleyenberg, K. C., Blasse, G.:* unpublished.
111. *Peacock, R. D.:* Strukture and Bonding, Springer Verlag, *22*, 83 (1975).
112. *de Hair, J. Th. W.* and *Blasse, G.:* J. Luminescence, 1976.

Received August 15, 1975

Vibrational Spectra of Oxo-, Thio-, and Selenometallates of Transition Elements in the Solid State

Achim Müller, Enrique J. Baran* and Roscoe O. Carter**

Institute of Chemistry, University of Dortmund. D 46 Dortmund, West Germany

Table of Contents

---

*) Guest Professor, Permanent Address: Facultad de Ciencias Exactas, Universidad Nacional de la Plata. La Plata, Argentina.

**) Alexander von Humboldt fellow from the U.S.A.

Problems in the Interpretation of the Vibrational Spectra of Chalcogenometallates in the Solid State

It is universally recognized that the vibrational spectrum of a crystalline salt is quite different from that of its solution. In recent years these differences have been systematically studied by a large number of research groups. The spectra have been recorded and carefully studied for most of the oxoanions of the main group elements. The proper assignments of the anion vibrations in the solid state are often based on the assignment of the spectrum for the solution. Despite the high symmetry and simple nature of the oxoanions of the transition metal elements improper vibrational assignments can still be found even in recently published books (*1—5*).

As the interpretation of the vibrational spectrum of solid state samples can provide information concerning the cation-anion interactions, it is of special interest to the chemist. In some cases information about the crystal structure can be obtained from the interpretation of spectra of the solid state when the substance possesses an anion of high symmetry.

An appreciation of the crystal field effect on the vibrations of the Bravais cell which is repeated to build the crystal is extremely important when interpreting the vibrational spectra of many substances, since in the presence of a crystal field influence the number of observed bands in the spectrum cannot be directly determined from the formula unit which goes to make up the unit cell. In other words, there is almost always a larger number of bands to account for when investigating solid state samples. The solid state effects often cause degenerate bands to split in the same degree as symmetric and antisymmetric stretching modes split.

In this review, the vibrational spectra of solid chalcogenometallates are presented and a critical discussion of the results given. Initially, measurements of powdered, crystalline samples with isolated ions or molecules are presented followed by single crystal Raman studies which are rarer. Additionally, a group of topics including the interpretation of Raman band intensities and widths, Resonance Raman spectra, the influence of pressure, temperature and sample preparation will be discussed.

1. General Background of the Theory of the Vibrational Spectra of the Solid State

1.1. General Considerations

The interpretation of the vibrational spectra of the solid state can be undertaken with several different orders of assumption. These methods of interpretation have been reviewed in detail elsewhere (*3—5* and *6—18*) and are given only in a brief summary here.

a) *Zero order approximation—orientated gas model.* In this assumption the vibrations of the isolated ions or molecules in the lattice are assigned according to the same point group and used for the "free" polyhedron in the gas phase or in solution. This approximation is very crude since it assumes firstly that the symmetry of the polyhedron in the crystal is the same as in an isolated state and secondly that no intermolecular interaction exists in the lattice.

b) *First order approximation — "site group analysis"* (19). The symmetry of the ions in the crystal is determined by the so called "site group", which on the one hand is a subgroup of the free ion and on the other hand is isomorphic to a subgroup of the space group. In this approximation only statistical crystal field effects are considered to influence the polyhedron. Intermolecular coupling between formula units are ignored. Since the "site symmetry" is in general lower than the symmetry of the "free" ion, degenerate vibrations can be split and those modes, which were forbidden, can become optically active. The use of the "site symmetry" group fails when the unit cell contains many formula units.

The various possible site symmetries for the different point groups have been tabulated in several places (3, 19—21). The necessary correlation tables between point and site groups, which are used in discussions of the examples in this text are assembled in the Appendix.

c) *Second order approximation — the factor group analysis or unit cell approxi-mation.* As suggested by *Bhagavantam* and *Venkatarayudu* (22), let us consider the internal vibrations of the ions or molecules as separate from the external modes for the purpose of the factor group analysis. In terms of the abstract group theory the factor group consists of the cosets of the invariant subgroup made up of the translation elements of the space group. The effect is that all the transla-tional elements of the space group form the identity element of the factor group [see also (23, 24)]. The factor group is always isomorphic with one of the 32 point groups. In general only $k=0$ phonons (wavevector $=k$) are observed in the spectra since they are found directly in the center of the Brillouin zone.

The factor group approximation considers the coupling with the formula units in the unit (Bravais) cell. In general the spectroscopic unit cell does not agree with the crystallographic cell since only the atoms which are not trans-lationally invariant are to be considered. Accordingly, the factor group analysis treats not only the statical crystal field effects but also dynamic effects. The reduction of the reducible representation of the point group which is isomorphous with the factor group on the basis of the cartesian displacement vectors of the atoms in the elemental cell, gives the number of optically active phonons.

In many cases it is possible to differentiate between the so called internal vibrations, those vibrations within the coordination polyhedron, and the external vibrations or lattice modes. The lattice modes can be of either the librational or translational type.

Methods for treating the factor group vibrations have been given by *Davydov* (25), as well as by *Bhagavantam* and *Venkatarayudu* (22). A simple analysis is possible through what is known as the "correlation method" (20, 26, 27) by which one is able to write the irreducible representations and thereby classify $k=0$ phonons directly and simply. The number of $k=0$ phonons is $3N$, where N is taken to be the number of atoms in the entire unit cell. However, there are only $3N$-3 optically active phonons because the acoustic vibrations have approximately zero frequency. The acoustic vibrations are easily determined since they have the same symmetry properties as the translations of the particular factor group being considered.

It is furthermore possible with the correlation method to divide the lattice vibrations into translational types and librational (rotational) types. The same results are obtained with this method as with that of *Bhagavantam* and *Venkatara-yudu*. *Adams* and *Newton* (*28, 29*) have recently published tables which make a simple factor group analysis possible [see also (*30*)].

In Section 6.11. a discussion of the measurement of the spectrum will be given. Here, however, it should be mentioned that a complete explanation of the spectrum of powdered samples with respect to the total vibrational assignment of all the observed bands is usually impossible. The number of single crystal studies is much smaller than the number of powder studies.

1.2. The Measurement and Interpretation of Single Crystal Raman Spectra

Single crystal Raman studies are very useful since in many cases a total assignment of all of the factor group fundamental vibrations is possible with this technique. However, it is necessary to orientate the crystal precisely. That in general is achieved by the determination of the optical indicatrix. The ability to assign the phonons directly stems from the fact that one records only certain phonons with a particular orientation, and different orientations can yield different phonons. As an aid for doing this one uses the Raman tensor found in the literature for the irreproducible representations of the various point groups. This tensor includes for each irreproducible representation only those elements, α_{ij}, whose differentials with respect to the normal coordinates are non-zero.

The symbolism of *Damen et al.* (*31*) will be used to describe the experimental conditions for a given spectrum. Hence, $i(kl)j$ means that polarized light from the i-direction illuminates the crystal and that the j-direction is to be observed. In so doing the electric vector of the inducing light is polarized in the k-direction. The other symbol contained within the brackets pertains to the polarization of the observed scattered radiation, then the quantities outside the brackets relate to the direction of propagation of light while those inside relate to the polarization of the electric vector of that light.

It would be best perhaps if this technique were illustrated by an example. In the case of single crystal measurements for a compound which crystallizes in the β-K_2SO_4 form (D_{2h}^{16}-P_{nma}; factor group D_{2h}), the Raman tensor for the A_g irreducible representation contains only the diagonal elements. Observation of the A_g phonons is possible only in the following orientations and conditions: $y(xx)z$, $x(yy)z$, and $x(zz)y$. This means that the polarizations are always mutually perpendicular to the two directions of light propagation.

In general it is not necessary to measure all possible configurations in order to identify all the phonons. However, it should be pointed out that frequently violations of the selection rules appear or that in a particular orientation forbidden phonons appear. These are believed to be caused by disorders in the single crystal.

2. Vibrational Spectra of the Chalcogenometallates

2.1. Oxometallates

In recent years numerous papers have been published on the vibrational properties of oxometallates. A short compilation of some of this data has been given in Refs. (32) and (33). Recently *Gonzalez-Vilchez* and *Griffith* (34) examined the infra-red and Raman spectra of a number of oxometallates. However, most of the values given there and the conclusions obtained from those values are not correct as we hope to demonstrate later. Various regularities as well as interdependences between spectroscopic behavior and other properties of these oxosubstances will be discussed [see also (35—37)].

There are a few general rules which can be formulated about the position of the stretching vibrations which are especially interesting and useful in making assignments (38, 39).

— For ions with central atoms belonging to the same group of the periodical system and possessing the same charge, the ratio $\nu_1(A_1)/\nu_3(F_2)$ increases with the increasing mass of the central atom (the ratio for MoO_4^{2-} is larger than that of CrO_4^{2-}).

— For a given central atom the ν_1/ν_3 ratio grows with the increasing charge (ν_1/ν_3 for MnO_4^{2-} or ν_1/ν_3 for MnO_4^-).

— For isoelectric ions where the central atom mass remains approximately constant, the ratio ν_1/ν_3 increases with the increasing charge of the anion (for example, the ν_1/ν_3 ratio is larger for CrO_4^{2-} than for MnO_4^-).

— The relationship ν_1/ν_3 remains approximately constant when ions of the same period and the same charge are compared (the ratio for CrO_4^{2-} is about the same as for MnO_4^{2-}). The application of these rules gives the proper assignment for other species such as CrO_4^{3-} and CoO_4^{4-} (40, 41), and the position of the $\nu_1(A_1)$ vibration for $MoSe_4^{2-}$ is verified (42). The applicability of these rules to chalcogenoanions of the main group elements has also been demonstrated (43).

Just recently we have formulated rules which are also effective in the region of the deformations (44).

— For tetrahedral ions or molecules of the type XY_4^{n-}, where X is a transition metal the $\nu_4(F_2)/\nu_2(E)$ frequency ratio changes from a value of less than one to a value greater than one with the increasing mass ratio M_x/M_y.

— In an isoelectronic row, the ν_4/ν_2 ratio increases with the increasing charge of the central atom (for example, the ratio for CrO_4^{2-} is greater than that for VO_4^{3-}).

— In one group of the Periodic Table, ν_4/ν_2 decreases with the increasing mass of the central atom (ν_4/ν_2 for CrO_4^{2-} is greater than that for MoO_4^{2-}).

2.1.1. Tetraoxometallates with d^0-Configuration

2.1.1.1. Titanates, Zirconates and Hafnates. Li_4TiO_4 (45) and Ba_2TiO_4 (46) are titanates for which the tetrahedral coordination of titanium IV with oxygen has been determined by X-ray analysis. The infrared spectrum of the latter com-

pounds has been determined in detail by *Tarte* (47—49). Bands at 755, 715, and 695 cm⁻¹ were found in the stretching region and bands at 374, 348, and 327 cm⁻¹ were found in the bending region. Considering only the "site symmetry" nine bands should appear in the infra-red spectrum, since the TiO_4^{4-} group occupies a C_2 site (46).

The first Raman study of Ba_2TiO_4 found a single strong and well defined band at 745 cm⁻¹ (50) which can certainly be assigned to the $\nu_1(A_1)$ vibration. As expected, ν_1 lies at higher frequency than ν_3 for this anion (38, 39). The IR and Raman frequencies of *Gonzalez-Vilchez* and *Griffith* (34) lie too high and yield an incorrect ν_1/ν_3 ratio as well as a too large a force constant when compared with Ref. (41). The correct normal vibrations for the TiO_4^{4-} ion are $\nu_1(A_1) = 750$ cm⁻¹, $\nu_2(E) \approx 360$ cm⁻¹, $\nu_3(F_2) = 705$ cm⁻¹ and $\nu_4 = 360$ cm⁻¹ (41). From the IR spectrum of Sr_2TiO_4 it is known that the oxygen atoms are octahedrally coordinated with the titanium (47, 49). However, there are very little data concerning ZrO_4^{4-} and HfO_4^{4-}, but the IR and Raman measurements indicate the possible existence of isolated T_d groups for Li_4ZrO_4 and Li_4HfO_4 (34).

2.1.1.2. *Vanadates, Niobates, and Tantalates.* The vibrational spectra of crystalline orthovanadates have recently been repeatedly investigated. Until now orthovanadates have been studied mainly in the presence of divalent cations. Infra-red data of a group of this type of compounds are collected in Table 1. The number of bands predicted ($Z_{theor.}$) for the stretching region by the "site symmetry analysis" are compared with the number of observed bands ($Z_{obs.}$).

Only the NaCl region has been measured for magnesium and calcium orthovanadates (51); thus no data are available for either of the bending vibrations. Previously the IR spectrum of $Ca_3(VO_4)_2$ had been discussed with incorrect crystallographic data. An X-ray structural analysis has only recently been reported (54) which shows that in the lattice crystallographically non-equivalent VO_4^{3-} groups are present. Also in the case of $Mg_3(VO_4)_2$ it was not previously possible to differentiate between two possible site symmetries, C_2 or C_s (51). A recently reported crystallographic study shows that the VO_4^{3-} ions sit on C_s sites (55).

The frequencies of the $\nu_1(A_1)$ vibrations have been obtained from the Raman spectra of $Sr_3(VO_4)_2$ and $Ba_3(VO_4)_2$. Values of 861 and 855 cm⁻¹ were measured for the strontium salt and 839 cm⁻¹ for the barium salt. These modes were too weak to be observed in the IR spectrum (52).

There is only little information available for orthovanadates with monovalent cations, as seen in Table 2. The IR spectra for $^6Li_3VO_4$ and $^7Li_3VO_4$ show clearly the bands associated with the VO_4^{3-} polyhedron. The high value of ν_1 for Tl_3VO_4 is certainly an error.

The IR spectra of orthovanadates with trivalent cations are also well known as seen in Table 3 (59, 60). In the IR spectra of the orthovanadates from Ce to Gd only one band less is observed as predicted and in no case is the ν_4 band observed to split. On the contrary, for $LaVO_4$ more bands appear than are predicted from "site symmetry" rules due to strong correlation field effects. For the same reasons the Raman spectrum of this compound is difficult to interpret (60).

IR and Raman single crystal measurements have been carried out for YVO_4 (61).

It is well known that vanadium can replace phosphorus or arsenic in the apatite lattice. Most of the orthovanadates which crystallize in this form have been investigated by IR and Raman spectroscopic techniques (62—64). The results of these studies are summarized in Table 4. The space group of all of these compounds is certainly C_{6h}^2 with the VO_4^{3-} ion lying on a C_s site. Therefore, one should always find a nine band spectrum. Phosphorus apatites behave similarly, as seen in Ref. (65).

Some orthovanadates which crystallize in the spodiosite lattice with the general formula M_2VO_4X have been spectroscopically investigated, but the spectra have not been fully discussed (66, 67).

So far only a few niobates and tantalates are known in which isolated tetrahedral NbO_4^{3-} or TaO_4^{3-} groups are present. Here an octahedral coordination is usually favored. Only for $YNbO_4$ and $YTaO_4$ have the measurements been carried out (68), the results of which are presented in Table 5.

2.1.1.3. *Chromates, Molybdates and Tungstates.* The vibrational spectra of chromates(VI) have repeatedly been studied. *Campbell* (69) measured a group of chromates in the spectral region from 250 to 1000 cm^{-1} without giving exact data concerning the band positions. K_2CrO_4 and $BaCrO_4$ have also been investigated by *Tarte* and *Nizet* (70), who compared these spectra with those of isostructural compounds. The IR spectra of chromates of singly and doubly charged cations have been measured in the NaCl region by *Baran* and *Aymonino* (71). Transition metal chromates (72, 73), trivalent lanthanide chromates (74) as well as chromates of hexamminecobalt(III), hexamminenickel(II), tetramminecopper(II), and tetramminezinc(II) (75, 76) have also been studied.

The Laser Raman spectra of various chromates have recently been repeatedly reported (77—82). A collection of the more relevant results for both the infra-red and Raman investigations are compiled in Table 6. Here it is evident that in general fewer bands have been observed than theoretically should appear. In particular the ν_2 deformation has in most cases not yet been observed in the IR and the ν_4 bands are often not split into all the theoretically predicted components.

In the case of Cs_2CrO_4 an explanation has been given for the absence of one of the ν_3 components (78), while a very strong correlation field effect is evident in the IR of $SrCrO_4$, although only six of the eight bands predicted by the factor group analysis for the stretching region have been observed (71). Silver, thallium and to a lesser extent, lead chromates all produce a clear shift of the band to a lower frequency (83) (see also Section 5). The presence of isolated CrO_4^{2-} ions cannot be inferred from the infra-red spectra for nickel, cadmium, and magnesium chromates (72). This is also evident in the spectra published by *Campbell* (69).

Molybdates and tungstates have likewise often been investigated, some of which have come out of our research group (84). These compounds can be subdivided according to crystal structure and these structures considered individually.

Na_2MoO_4 and Na_2WO_4 crystallize in the cubic space group O_h^7 ($Z = 8$) with a T_d site symmetry for the anions (85). Hence, the Raman spectrum should have four bands and the IR two according to site group analysis. Despite numerous attempts (86—91), the assignment of the bending region has only recently been convincingly explained. The assignment of $\nu_4 > \nu_2$ given by *Preudhomme* and

Tarte (*89*) for the Raman effect has been confirmed by new measurements (*84*). Our values from the Raman (*84*) are given in Table 7 for both compounds.

Monoclinic alkalimolybdates and -tungstates have been obtained (*92, 93*) and studied spectroscopically by *Caillet* and *Saumagne* (*88*). These salts have the space group C_{2h}^3, $Z=4$ and anionic site symmetry C_s. The IR and Raman data (*84*), given in Table 8 clearly show that the ν_3 band does not always split into the three expected components. The assignment of the deformation band can be made with confidence from the relative IR and Raman intensities. The ν_4 band is always more intense in the IR spectrum than ν_2, while the reverse is true in the Raman.

The heavy alkali molybdates and -tungstates are known to exist in an ortho-rhombic modification as well, having the space group D_{2h}^{16}, ($Z=4$) (*85*). The IR and Raman data (*84*), reproduced in Table 9, show clearly that ν_1 and ν_2 appear in the spectra for all these substances and that ν_3 and ν_4 are split into three bands in the Raman effect. These comply for the most part with the simple site symmetry treatment where the anion has C_s symmetry.

Most divalent cation molybdates and tungstates crystallize in the scheelite lattice with the space group C_{4h}^6, $Z=4$. These compounds have been studied using the usual IR and Raman techniques (*94—96*). Single crystal studies (*96—98*) and samples with pure isotopes of Mo and Ca (*99—101*) have been reported. Additionally, recent information shows that ν_4 lies above ν_2 (*84*). It should be pointed out here that one cannot speak about purely internal vibrations since coupling with lattice vibrations is not negligible, as seen in Ref. (*101*).

A large collection of molybdates and tungstates with differing known and un-known structures have been studied by *Clark* and *Doyle* (*95*) as well as by *Schwing-Weill* and *Arnaud-Neu* (*102*). However, no clear assignments have been reported in this work and no detailed discussion has been given for the numerous spectra. In some cases studies indicate the presence of MoO_6 and WO_6 groups in the crystal lattice and also distorted tetrahedra as well, as seen in Refs. (*73*), (*95*), and (*103*). In addition, IR and Raman spectra of transition metal tungstates with the wolf-ramite-structure have recently been measured (*104*).

2.1.1.4. *Permanganates, Pertechnetates, and Perrhenates.* Numerous reports concerning vibrational studies of permanganates exist. Alkali permanganates were originally measured and assigned by *Rocchiccioli* (*105*), although some of these assignments required changing (*106*). Other investigations resulted in correct assignments (*81, 107, 108*). Apart from the alkali permanganates there have been other studies of monovalent cation permanganates, such as the silver- (*107, 108*), ammonium- (*107, 109*), tetraphenylarsonium- and tetraphenylphosphonium- (*110*) salts. The data for the singly positive cations are assembled in Table 10, while those for the doubly positive cations are in Table 11.

The internal vibrations of the MnO_4^- ion seem to be influenced less by the cations than other metal-oxygen vibrations [see(*108*)]. For example, the iso-typical potassium-, rubidium-, cesium-, and ammonium permanganates have practically the same ν_1 and ν_3 frequencies. The difference observed in the case of $AgMnO_4$ is explained in Ref. (*83*). By the large cations, such as tetraphenylar-sonium and tetraphenylphosphonium, the ν_3 band is very sharp and well defined. Since these ν_3 bands are not split as expected it can be concluded that the anion is strongly shielded by these large organic type cations. Thus, the expected preservation of the T_d symmetry develops to a much lower degree than usual.

Barium- and strontium permanganate show clearly the effects of the correlation field effect by well split ν_3 bands. However, the ν_3 band never splits in the predicted way for salts of hexaaquo cations. This again is most likely due to a decreased distortion of the T_d units because of screening by the $[M(H_2O)_6]^{2+}$ cations.

Some other permanganates of unknown structure have been reported (111). The results of studies of complex cation permanganates which have only recently been studied (75, 76, 113), are given in Table 12. The IR spectra of tetramminezinc and tetramminecadmium permanganate should easily be interpreted from the site group selection rules, since the structural data is available (76). Nevertheless, the ν_3 bands are clearly split contrary to prediction. Either the effects of the correlation field are seen or the T_d^2 space group which describes the actual structure very well when considering the X-ray scattering intensities calculated for $Z=4$, cannot be used to interpret the spectra through the site symmetry. This last explanation is possible because these substances may possess a larger super-structure (76, 114).

The vibrational spectra of pertechnetates have only lately been widely studied. *Busey* and *Keller* (86) have reported the original IR and Raman spectra of crystalline $KTcO_4$. The measurements were later completed (85, 115). Additional salts have also been measured (107). All these data are collected in Table 13. The IR data for tetraphenylarsonium and tetraphenylphosphonium pertechnetate can be found in Ref. (115). The $\nu_1(A_1)$ stretch in pertechnetates has been observed in the Raman effect and found to lie at 910 cm^{-1} for $KTcO_4$ (84).

Numerous crystalline perrhenates have been studied in the last few years (84, 86, 115–120, 107). The IR data for substances of known structures are given in Table 14. The assignment has been given according to our most recent information (44, 121). The ν_1 vibration for these compounds has only been observed in the Raman effect and is found at 958 cm^{-1} for $NaReO_4$ (84), at 965 cm^{-1} for $KReO_4$ (120) and at 942 cm^{-1} for $AgReO_4$ (120). The Raman data for other perrhenates of unknown structure follow in Table 15. Most of these spectra show a clear correlation field splitting as ν_3 is split into four to five components and sometimes ν_1 is also split. An exact assignment of ν_2 and ν_4 in most cases appears difficult, so that we have given both vibrations together in Table 15.

Additionally, the tetrabutylammonium- and tetraphenylarsoniumperrhenate have been spectroscopically studied. Exactly as seen in the case of the pertechnetates and permanganates, the ν_3 bands are sharp and well defined. The perrhenates of different tetrammine- and hexammine-cations (75, 76) and rare earth cations (121 a) have also been reported.

2.1.1.5. *Ruthenium and Osmium Tetroxide*. OsO_4 crystallizes in the space group C_{2h}^6 with four molecules per unit cell and lies on a site with C_2 symmetry. The crystal structure of RuO_4 is known, but the similarity of the vibrational spectra of these two substances suggests similar structures. No single crystal data is available for either compound and only a few measurements on the solids have been carried out (123, 124). The two stretching vibrations are clearly assigned in both cases. On the other hand, the assignment of the bending region has until recently remained somewhat unclear. A comparison of the spectra of solid OsO_4 with that of the gas phase shows no great difference as seen from Refs. (124) and (125). This is as expected since the crystal field effects and interactions are smaller in molecular crystals than in ionic crystals. In this case no complete IR spectrum has been reported for RuO_4, but the Raman spectra have been recorded for the solid (123), liquid and gaseous states (126—128). The Raman data for both tetroxides in the solid phase, given in Table 16, are assigned as follows: $v_2 > v_4$ for OsO_4 and $v_4 > v_2$ for RuO_4.

2.1.2. Tetraoxometallates with d^n-Configuration

The vibrational studies of oxygen containing compounds with d^n-configurations have only lately been investigated. Many compounds in which the metal atom exhibits unusual or infrequently occuring oxidation states, such as Mn(V), Cr(V), Co(IV), belong to this group [see Refs. (129—131)].

2.1.2.1. *Tetraoxometallates with d^1-Configuration.* Anions such as VO_4^{4-}, CrO_4^{3-}, MnO_4^{2-}, and RuO_4^{-} fall into this category. Very little is known about VO_4^{4-} in which vanadium has an oxidation number of four. Ba_2VO_4 and Mg_2VO_4 were studied in the afore mentioned work by *Gonzalez-Vilchez* and *Griffith* (34). The reported values follow the rules outlined in the beginning, since $v_1 = 818$ cm^{-1} and $v_3 = 780$ cm^{-1}, the ratio of v_1/v_3 gives 1.05 which is, as expected, similar to that found for TiO_4^{4-} and CoO_4^{4-} (41).

Compounds containing CrO_4^{3-} with chromium(V) have been repeatedly investigated. *Guerchais et al.* (132) have studied Li_3CrO_4 and $Ba_3(CrO_4)_2$ for the first time. The latter substance was simultaneously measured by *Baran* and *Aymonino* (35) and also somewhat later by *Müller* and co-workers (133.) Additionally, $Ca_3(CrO_4)_2$ and $Sr_3(CrO_4)_2$ have been studied in the infra-red. $Ca_3(CrO_4)_2$ more than likely possesses the same structure as $Ca_3(VO_4)_2$, so that crystallographically non-equivalent CrO_4^{3-} groups probably exist in the lattice. An improper assignment has been given by *Doyle* and *Eddy* (134) which was later corrected in Ref. (40) and (135). Barium and strontium chromate(V) have been measured recently in the infra-red down to 200 cm^{-1} by *Tarte* and *Thelen* (136). All the reported data for chromate(V) compounds are collected in Table 17.

In the stretching region of $Sr_3(CrO_4)_2$ and $Ba_3(CrO_4)_2$ only two strong bands appear, as by the isostructural orthovanadates given in Section 2.1.1.2. These two bands are here assigned to the split v_3 mode, as expected theoretically. The v_1 band is too weak to be identified. Only one band by 345 cm^{-1} was found by *Tarte* and *Thelen* (136) in the bending region for $Sr_3(CrO_4)_2$, which is assigned with certainty to the v_4 vibration. For $Ba_3(CrO_4)_2$ two bands were found which can be treated as either two components of v_4 or one from v_4 and one from the v_2 band.

The compound with the formula Ca_2CrO_4Cl, having spodiosite structure, has been studied in the infra-red (137) but bands in the stretching region were incorrectly assigned.

Manganate(VI) has also been the subject of research as in, for example, $BaMnO_4$ by *Guerchais* and co-workers (132) as well as *Baran* and *Aymonino* (35). This substance has a baryte structure with the MnO_4^{2-} ions laying on lattice points with C_s symmetry. In the stretching vibration region all four expected bands are observed at 888, 847, 826 cm⁻¹ for the ν_3 components and 815 cm⁻¹ for ν_1 (35). Alkalimanganates of the type M_2MnO_4 have been studied by infra-red absorption in the NaCl region (105, 138). *Rocchiccioli* (105) observed only an unsplit ν_3 band, whereas *Baran* (138) measured in every case a clear ν_1 band and some splitting of the ν_3 band as well, as seen in Table 18. K_2MnO_4 crystallizes in the β-K_2SO_4 form (139), so that in the NaCl region four bands are expected. This substance has a similar spectrum to that of K_2SO_4 (70).

Spectra of $NaRuO_4$ and $KRuO_4$ have been communicated (34, 140). $KRuO_4$ crystallizes in the scheelite lattice leading to an expected twofold splitting of both the ν_3 and ν_4 bands.

2.1.2.2. *Tetraoxoanions with d^2-Configuration.* CrO_4^{4-}, MnO_4^{3-}, FeO_4^{2-}, and RuO_4^{2-} belong to this class of compounds. The values given by *Gonzalez-Vilchez* and *Griffith* (34) for Ba_2CrO_4 seem to be incorrect as well as the values of *Campbell* (141) for the main bands of this compound. The samples which were used in these studies may have possibly contained large quantities of chromate(VI). Moreover, for CrO_4^{4-} ν_1 must be greater than ν_3. Certainly then the values reported for MnO_4^{4-} and WO_4^{4-} are also incorrect.

For the manganate(V), $Ba_3(MnO_4)_2$ has been especially widely investigated (35, 132, 136). Only two bands at 821 and 757 cm⁻¹ are observed (35) in the stretching vibration region similar to those seen for the corresponding isostructural orthovanadate and chromate (V). These two bands are ascribed to the two components predicted for ν_3. In the bending vibration region again only two bands are found at 342 and 316 cm⁻¹ (136) which are attributed to the expected splitting of ν_4. In this case ν_1 and ν_2 have never been located. The values reported by *Gonzalez-Vilchez* and *Griffith* (34) for Cs_3MnO_4 and K_3MnO_4 are unreasonable for ν_3.

The potassium and barium ferrates have not been discussed frequently (70, 142). The IR data for these two compounds as well as for Cs_2FeO_4 are reported in Table 19. Rb_2FeO_4, which is isostructural with the corresponding potassium and cesium salts and displays a similar spectrum, but exact frequency data are not available (142). The IR spectra of $BaFeO_4$, $SrFeO_4$ and Na_2FeO_4 have been presented in full by *Bécarud* (143), however, some of the results given are dubious. For example, for $SrFeO_4$ the band at 432 cm⁻¹ appears too high to be one of the ν_4 components. Furthermore, the ν_4 component is missing completely for $BaFeO_4$ even though it was clearly present in the spectrum reported by *Tarte* and *Nizet* (70).

For the ruthenium(VI) oxoanion only $BaRuO_4$ has been studied in the solid state (140) with the following results: $\nu_1 = 856$, ν_2 unobserved, $\nu_3 = 812$ and $\nu_4 = 330$ cm⁻¹. The solution spectrum for K_2RuO_4 has also been measured (34, 140). No corresponding oxoanion of osmium is known as $K_2OsO_4 \cdot 2\,H_2O$ shows no evidence of OsO_4^{2-} ions. This substance must be formulated as $[OsO_2(OH)_4]^{2-}$ (144, 145).

2.1.2.3. *Tetraoxoanions with d^3-Configuration.* Here only one example is known, namely ferrate(V), FeO_4^{3-}. The only published data for K_3FeO_4 (34) is certainly incorrect. K_3FeO_4 is very unstable and hence difficult to measure without decomposition.

The samples measured in Ref. (34) are probably strongly contaminated by K_2FeO_4. Some measurements of K_3FeO_4 (146) indicate that the ν_3 band lies at approximately 760 cm^{-1}. Accordingly, the corresponding ν_1 should appear around 780—800 cm^{-1}. Hence, the ν_1 value given in Ref. (34) could be approximately correct, however, the same cannot be inferred for the ν_3 value. Furthermore, $\nu_1 > \nu_3$ must be valid as well for FeO_4^{3-}.

2.1.2.4. Tetraoxoanions with d^4-Configuration. Of the two anions belonging to this group, CoO_4^{3-} and FeO_4^{4-}, practically nothing is known. K_4FeO_4 has been described (147) but not studied spectroscopically. This substance is, as was K_3FeO_4, very difficult to obtain pure and is very unstable.

The barium salt of ferrate(IV) has also been studied (34), but the results are doubtful. The ν_3 value is extraordinarily high for a Fe(IV) compound and leads, furthermore, to an unreasonable force constant as seen in Ref. (33), as well as to an incorrect ν_1/ν_3 ratio.

2.1.2.5. Tetraoxoanions with d^5-Configuration. The only known example, CoO_4^{4-}, has only lately been investigated in Ba_2CoO_4 (41). The values given by *Gonzalez-Vilchez* and *Griffith* (34) are apparently incorrect. Correct values for the CoO_4^{4-} normal vibrations are: $\nu_1 = 670$, $\nu_2 \approx 320$, $\nu_3 = 633$, and $\nu_4 = 320$ cm^{-1} (41).

2.1.3. Pentaoxometallates

Up to now very little reliable data have been given in the literature for pentaoxo-metallates. According to data given by *Griffith* (148), pentaoxometallates such as K_3OsO_5 and K_3ReO_5 do not contain isolated MO_5 groups, although one pentaoxo-metallate containing tetragonal-pyramidal MO_5^{n-} groups has been identified by X-ray structural investigations ($148a$).

2.1.4. Hexaoxometallates

Different hexaoxometallates have been studied by *Hauck* (149—152) and by *Corsmit et al.* (153). *Baran* and *Müller* (154) have clearly found by means of IR spectra that $Ba_5(ReO_6)_2$ contains isolated ReO_6^{5-} octahedra. The known data are collected in Table 20 for different hexaoxometallates; however, most of these measurements are incomplete. The IR data for ν_3 and ν_4 are only partly correct, since these bands are very broad for some of the lithium salts so that it is often difficult to differentiate between ν_3, ν_4, and the ν(Li—O) vibrations (155). In most cases it is doubtful whether the IR bands forbidden for the free ion, ν_1, ν_2, ν_5 are correctly assigned. Surprisingly enough ν_3 and ν_4 do not split in most of the IR spectra even though there is enough perturbation to allow the appearance of normally forbidden transitions. Several lithium hexaoxometallates have been investigated in the Raman effect (155).

2.2. Thio- and Selenometallates

2.2.1. Tetrathiomolybdates and -tungstates

The IR and Raman frequencies for the alkali tetrathiomolybdates and -tungstates (156) follow in Table 21. All these compounds crystallize in the β-K_2SO_4 lattice, space group D_{2h}^{16} with $Z = 4$ having anionic site symmetry C_s (157). The number of bands observed in the IR and Raman spectra is usually less than expected for the site group or factor group analysis. $\nu_1(A_1)$ can usually be seen in the IR spectrum while the splitting of the $\nu_3(F_2)$ band is clearer in the Raman effect. The exact assignment of the ν_2 and ν_4 modes is difficult as they are accidentally degenerate in the solution spectra. However, according to recent reports, the intensity of ν_2 should be greater than that of ν_4 in the Raman effect (44, 121). The most probable assignments are therefore also given in Table 21. No effect is apparent that is clearly dependent on cation. The thallium(I) salts of these anions have also been investigated (158).

2.2.2. Other Tetrathio- and Tetraselenometallates

The solid state spectral data for thiovanadates, -niobates, -tantalates, and -perrhenates as well as the tetraselenometallates are given in Table 22. All these compounds are summarized here since no systematic study of the influence of the cations has yet been undertaken. The selenometallates of the group VIB (given in Table 22), like the corresponding thiocompounds crystallize in the β-K_2SO_4 form (160), while the corresponding group VB metallates crystallize in a cubic structure, space group T_d^2 (159). The IR spectra have been measured for ammonium tetraselenomolybdate and -tungstate (163).

2.2.3. The Mixed Thioselenometallates of the Type $A_2MS_xSe_{4-x}$

Cesium salts of these compounds crystallize in the β-K_2SO_4 form with statistical orientation of the anions. Therefore, one observes no splitting of the degenerate vibrations (see Table 23) similar to anions with C_{3v} symmetry such as $KCrO_3F$ and $KOsO_3N$ (164). The structures of these compounds are discussed in Ref. (160). Due to the method of preparation, the solubilities and the nature of the structure, i.e. the possibility of mixed crystal formation, the x-values are not exactly integers (165).

2.2.4. Trithiomolybdates and -tungstates

The vibrational frequencies for compounds of these anions with various cations are given in Table 24. The splitting of the degenerate vibration is only occasionally observed. It is not possible to differentiate between the $\delta_{sym}(MS_3)$ and $\delta_{asym}(MS_3)$. The same has been found for the solution spectra. All these compounds crystallize in the space group $D_{2h}^{16}-P_{nma}$ (166, 167).

Only recently compounds of the type $K_3(MOS_3)X$ [$M=Mo, W; X=Cl, Br$] have been prepared (168, 169). Of particular interest here is the splitting of the $\varrho(MS_3)$ vibration (169a).

2.2.5. Other Thio- and Seleno-Oxocompounds with Anions of C_{3v} or C_{2v} Symmetry

In addition to the trithio-compounds discussed in Section 2.2.4., the corresponding triseleno salts are also known. The IR and Raman data for Cs_2MoOSe_3 and

Cs_2WOSe_3 have been reproduced in Table 25; the data for salts with ions of the type $MO_2X_2^{2-}$ (M = Mo, W and X = S, Se) follow in Table 26 (158, 170). Of particular interest is the marked lowering of ν(M—O) frequencies for the ammonium salt due to hydrogen bridge formation (171). There is no expected site group splitting for these ions. The crystal structure and chemistry of these solids are discussed in Refs. (171) and (172).

2.2.6. Thallium and Copper Chalcogenometallates

The data for these compounds can be found in Tables 27 and 28 (173). The frequencies for Tl_3MX_4, where M is V, Nb or Ta and X is S or Se, can readily be taken from Table 22. In every case the chalcogen-metal vibrations are markedly decreased. The compounds such as Cu_3MX_4 where M is V, Nb or Ta and X is S or Se form a three dimensional crosslinked network due to strong Cu—X bonding. The coppermetallates crystallize in the sulvanite form.

Recently compounds of the type $Cu_3MS_xSe_{4-x}$ where M is Nb or Ta have been prepared and found to form a continuous series of mixed crystals (160, 174). The X-ray investigations indicated the existance of a sulvanite structure (160). The vibrational spectra proved to be very complicated.

3. Vibrational Spectra of Various Anions in Compounds with the same Structure

If the IR spectra of a group of crystalline substances with the same structure are studied, then certain characteristics of the spectra can be found which are always present in that crystal form. Here we list a few typical examples of this statement.

3.1. The ß-K_2SO_4 Structure

For compounds of this group, such as K_2CrO_4, K_2FeO_4 and K_2MnO_4, the ν_3 region usually does not have the three expected components; one often observes only one broad strong band with one or two poorly split shoulders. The ν_1 band is clearly recognizable in all of these compounds, even though it tends to be weak, since it is always well defined and sharp.

3.2. The Baryte Lattice

The three ν_3 components are clearly separated in this group of substances which include $BaCrO_4$, $BaMnO_4$ and $BaFeO_4$. The ν_1 band is always relatively intense while the ν_4 band is in most cases split into two components but the ν_2 mode is generally not observed.

3.3. The $Ba_3(PO_4)_2$ Structure

The following substances among others belong to this group: $Ba_3(VO_4)_2$, $Sr_3(VO_4)_2$, $Ba_3(CrO_4)_2$, $Sr_3(CrO_4)_2$, and $Ba_3(MnO_4)_2$. For all these compounds only two IR bands are seen in the stretching region, which are assigned to the A and E components of the ν_3 mode. The ν_1 vibration, although it is infra-red active appears to be so weak that it is masked by the ν_3 components and is thus not observable. For both of the above mentioned orthovanadates, the ν_1 vibration was found to lie between the two ν_3 components (175). This behavior has not been demonstrated for any of the other compounds.

3.4. The Scheelite Structure

The compounds belonging to this structural group have spectra in the stretching region which are very similar. The fine structure of the ν_3 band in the infra-red absorption spectra for $CaMoO_4$ and $CaWO_4$ on one hand and $KTcO_4$ and $KReO_4$ on the other is very similar.

3.5. The Zircon Structure ($ZrSiO_4$)

All the orthovanadates with zircon structure and formula $LnVO_4$ where $Ln = Ce$, Pr, Nd, Sm, Eu, and Gd, which have been studied up to now, have very similar spectra. The ν_3 vibration which is predicted to split into two bands from the site symmetry treatment appears in the IR as one strong, broad band with a weak shoulder. On the other hand, ν_4 appears as a sharp band. $CaCrO_4$, which has the same structure, possesses a very similar spectrum.

3.6. The Mg ($ClO_4)_2 \cdot 6H_2O$ Structure

Permanganates, which crystallize in this structural form, i.e. Zn-, Cd-, Mg-, and Ni-permanganate have only one fairly symmetric band for ν_3 in the IR. ν_1 has been found to be less clearly apparent than in other permanganates, for example those having β-K_2SO_4 structure. It seems that the large $[X(H_2O)_6]^{2+}$ cation lessens the perturbation of the lattice and screens the MnO_4^- ion.

4. Spectra of Compounds with Metal Isotopes

Tarte and co-workers measured the first metal isotope shifts for oxo- compounds in order to obtain a definite vibrational assignment of the solid (101). Among other isotopic pairs, they worked with the following: [24/26]Mg, [50/54]Cr, [58/62]Ni, [64/68]Zn, and [70/76]Ge.

With [40/44]Ca- and [92/100]Mo- data the differentiation of the translational and librational modes was obtained for the molybdates and tungstates with scheelite structure (98, 100, 101). Tables 29 and 30 reproduce these results. In contrast to early findings, the lower frequency bands were found to be librations and not translations. Furthermore, the band associated with ν_4 for $CaMoO_4$ and $CaWO_4$ cannot be expressed as a pure deformational vibration of the MO_4^{2-} group, but this mode is coupled with the translational vibrations, as indicated in Table 29. The relationship between the translational vibration, E, and the square root of the mass of the cation for compounds of the type AMO_4 ($A = Ca$, Sr, Ba, Pb; $M = Mo$, W) was determined to be linear (100).

Similarly, by using this method for a series of substances with spinel structure some confusion was definitely removed. In this way, for example, the assignment in the deformation region for Na_2WO_4 and Na_2MoO_4 (89, 90) has been clarified. Measurements have been carried out on the $^{92/100}MO$ isotopes for molybdates as well (90, 176), and theoretical calculations have been done for the intensity ratio for ν_2 and ν_4 in the Raman (44). Theoretically $I\ (\nu_2)$ should be greater than $I\ (\nu_4)$ for oxoanions in solution. This should also be true for cases where there is only a small degree of coupling with the lattice vibrations for spectra of solid state samples, as for example Na_2MoO_4 (121).

The interpretation of the lattice vibrations for scheelite type molybdates or tungstates with relatively light cations, Ca or Sr, has indicated that the lowest translational vibrations are produced by Mo—Mo or W—W motions respectively, while those at higher frequency are from cation-cation motions (98). This has not been found, however, in the case of the barium or lead compounds. The librational frequencies have been found to decrease linearly with the ionic radius of the cation for AMO_4 type compounds, where $A = Ca$, Sr, Ba, or Pb and $M = Mo$ or W (98).

Further investigations by *Tarte* and co-workers have shown in the case of II—IV germanates, I—II—V vanadates, and I—VI molybdates and tungstates that the bands at higher wavenumbers in every case must be attributed to the vibrations of the tetrahedral grouping, (89) *i.e.* the metal oxygen stretch for the metal in the highest oxidation state. The isotopic data show that vibrations at middle and lower wavenumbers are more complex in nature. The above is always true in our opinion where either the polarization effects of the "cations" are small, like Na, or when the oxidation state of the "central atom" of the anion is high (V, Mo).

The validity of the Product Rule for the F_2 class is dependent on the isolation of the anion in its lattice. For a XO_4 ion for which only X is being substituted, the Redlich-Teller Product Rule can be written as follows:

$$\frac{\nu_r \cdot \nu_d}{\nu_r^i \cdot \nu_d^i} = \frac{m_x^i(m_x + 64)}{m_x(m_x^i + 64)}$$

where ν_r and ν_d are antisymmetric, stretch and bend respectively and m_x is the mass of the central atom X. A comparison of the experimental ratios with the calculated ones is given for several compounds in Table 31. (89) It is obvious that the Product Rule is fulfilled for the molybdates. This is expected since the bonding between the octahedrally surrounded cation, either Na or Ag, and the oxygen atoms is certainly small in comparison with the bonding between Mo(VI) and oxygen. This is not the case for the germanate where the lower value of the formal charge of the "central atom" causes a weakening of the bonding in the "tetrahedral anion". In these cases, the "cations" participate in the internal vibrations of the GeO_4-ions (89).

Investigations of the vibrational spectra of spinels of the II—III type (177) and $LiYX_4O_8$ type (178), which are not built up of discrete units, have also been reported. In normal cubic II—III spinels four bands are observed in the IR spectrum, as predicted by theory (177). The analysis of the spectra for solid solutions as well as isotopic data indicate that ν_1 and ν_2 correspond roughly to the vibrations of the octahedrally co-ordinated trivalent cation. In comparison, ν_3 and ν_4 correspond to vibrations for which octahedral and tetrahedral groups must be considered (177, 179).

From the above results one sees that complete information about the assignment of internal and external vibrations can be obtained with the help of metal isotope data.

5. The Influence of Cations on Internal Vibrations

For a collection of similar substances with the same anions but different cations, a slight but significant shift can be seen in the anion vibrational frequencies. As the crystal lattices are identical, these observed shifts depend on the nature of the individual cations. As yet, no unifying explanation has successfully described the dependence for all of these observed facts. One cannot yet say with certainty which property of the cations is the controlling factor. Formal charge, effective nuclear charge, electronegativity, and ion radius are just a few of the factors which have been considered. To what degree the individual factors by themselves or in conjunction with another operate in generating the observed behaviour, cannot in general be exactly determined.

Guerchais and co-workers (180) have shown that there is a linear relationship between the cation radius and the symmetric stretching frequency of the sulfate group for alkalisulfates with β-K_2SO_4 structure, i.e. the frequency decreases with increasing radius. On the other hand, *Adler* and *Kerr* (181—183) have found a dependence between not only the radius, but also the mass of the cation and the vibrational frequencies in a set of investigations on a group of naturally occurring carbonates and sulfates. *Weir* and *Lippincott* (184) and later *Adler* (185) and *Laperches* and *Tarte* (186) have shown, however, that there is no significant influence of the mass of the cation on the internal vibrations of the anions.

The consideration of these problems (138, 187) lead us to the conclusion that the polarizing power of the cations as measured by the effective nuclear charge alone, is most probably the influential parameter necessary to understand these effects. The polarizing power given otherwise as formal charge/radius or formal charge/radius^2 or even better, effective nuclear charge/radius, cannot explain these effects as well.

In general, the anion vibrational frequencies move to longer wavelengths with increasing effective nuclear charge of the cations. This seems in principle to be reasonable, because the larger the effective nuclear charge of the cation, the stronger is the expected interaction between the cation and the anion. These interactions lead naturally to a weakening of the bond in the anion.

In many cases the relationships have not been so easily interpreted. For many salts, having transition metal cations, no unequivocal trend can be found (138). The electronic configuration of the cation has also been found to have an effect in this direction. However, the exact opposite trend has been reported for other compounds such as vateritetype lanthanide borates, the vibrational frequencies increase with increasing effective nuclear charge of the cation (186, 187). All this goes to indicate how complicated this problem is.

Silver and thallium(I) as well as copper(I), lead(II) and mercury(II), in most cases, show especially strong effects with respect to anionic vibrations. One must assume then that the interactions are especially strong in these cases. A systematic compilation of results for numerous anions crystallized with Ag(I) and Tl(I) cations has been published (83). Both cations usually show drastic deviations from the plot of effective nuclear charge versus vibrational frequency (see 83, 138). These startling effects are seen not only with oxoanions but also with thio- and selenoanions as seen in Refs. (188—190). It can be assumed then that some form of a covalent bond exists between the silver or thallium atoms and the chalcogeno atoms. The bonds are formed from p-orbitals from oxygen, sulphur or selenium, respectively, which for most anions have p_π character. After such co-participation of the chalcogenoatoms the M—O (or M—S or M—Se) bonds should be weaker. Hence, the p-orbitals of the chalcogens are no longer fully free to participate in the π-bonding within the anion, thereby weakening the bonding in the anion.

When a cation contains hydrogen there is always the possibility of hydrogen bridge formation between cation and anion which can again weaken the bonds in the anion. This effect is especially strong for $(NH_4)_2MoO_2S_2$ and $(NH_4)_2WO_2S_2$ (171). In addition, in ammine complexes such as $[Co(NH_3)_6]^{3+}$, $[Cu(NH_3)_4]^{2+}$, and $[Ni(NH_3)_6]^{2+}$, similar interaction can occur through hydrogen bridges. It is interesting to note that the strongest of such effects has been observed for dithio-oxoanions (75, 76). On the other hand some vibrations of these complex cations, especially the rocking vibrations, are dependent on the anion (191). These data clearly show the presence of anion-cation interactions.

Large cations, such as the afore mentioned ammine complexes or organic cations like tetraphenylarsonium or tetrabutylammonium, have practically no influence on the internal vibrations of the anion. On the one hand the classical polarizing effect of the cation would be decreased because of the greater radius; on the other hand, the large volume of the cation screens the anion so that all the lattice interactions would be decreased. Usually, in such cases, sharp and very clear anion bands are found, indicating screening (110, 115).

Cation related effects are frequently found and discussed in spectroscopic studies for anions containing transition metals. Some typical examples are given here:

Vanadates: For the isostructural strontium and barium ortho-vanadates the V—O vibrational frequencies decrease with increasing effective nuclear charge [51]. Similar orthovanadates with the apatite structure have for a given halogen ion decreasing frequencies of the VO_4^{3-} ion in the order Ca > Sr > Ba [(62) and Table 4]. However, it is impossible to see any trend towards the trivalent lanthanide cations (59). The IR bands are so wide in this case that small changes cannot be measured.

Chromates: For $Sr_3(CrO_4)_2$ and $Ba_3(CrO_4)_2$, as was true for the corresponding vanadates, a decrease in the frequencies was seen when going from Sr to Ba (71). The frequency of the ν_1 vibration was found to decrease for the following compounds as indicated:

$$K_2CrO_4 > Rb_2CrO_4 > Cs_2CrO_4 \ (78).$$

Similarly, the position of the ν_3 components moved correspondingly to lower frequency (see again Table 6). The two ν_3 components of the lowest frequency band draw together, so that they are 12 cm^{-1} apart in K_2CrO_4, 8 cm^{-1} in Rb_2CrO_4 and apparently fall on top of one another in the Cs salt. These examples demonstrate that it is important to compare isostructural compounds in order to obtain an uncontestable spectral assignment. Without such a comparison one could not explain the absence of one band in the Cs_2CrO_4 spectrum.

Manganates: The isostructural series K_2MnO_4, Rb_2MnO_4 and Cs_2MnO_4 show a regular decreasing trend for the ν_1 vibration (138), as can be seen from the data in Table 18. More interesting, however, is the behavior of the permanganate. With the single exception of $AgMnO_4$, which has been previously described, the vibrational spectra of MnO_4^- show almost no dependence on cation (108, 138). This points to an extremely strong manganese-oxygen bond. Interestingly, MnO_4^- has been found to have the largest f_{M-O} force constant for all tetraoxo-anions of the first transition series (192). This explanation is, however, not completely satisfactory as the perrhenate anion, which has an even higher f_{M-O} force constant, clearly shows the influence of cations. This peculiarity of the permanganate ions certainly deserves further investigation in order to clarify this situation.

Perrhenates: The presence of a systematic shift in the vibrational frequencies for the perrhenate ion has been demonstrated for a large number of cations (120). The alkali cations from Na to Cs, were shown to increase the vibrational frequencies as a function of increasing effective nuclear charge. This is just the opposite effect discussed for the previous compounds. *Ulbricht* and *Kriegsmann* (120) attributed this to the decrease of the electronegativity of the cation in the direction Na to Cs or, even better, a decrease in the electronegativity/radius^2 *cation* quotient. Divalent salts of perrhenate, however, display the expected behavior, *i.e.* decreasing frequencies of vibration with increasing effective nuclear charge of the cation. This alternation of behavior in going from mono to divalent cations indicates that interactions between anion and cation have become significant in the divalent cation case (120).

Finally we should mention that the cationic influences are greater on oxo-anion than on pure or mixed thio- or selenoanions.

6. Special Problems and Measuring Techniques

6.1. Raman Measurements on Single Crystals

We pointed out in the beginning that the direct assignment of phonons can be made by means of measurements of oriented single crystals. Thus, more phonons have been found for YVO_4 single crystals than were previously measured (193). The spectra of single crystals of alkali chromates have been measured by two groups (79, 194) with good agreement for the internal vibrations. Table 32 contains the correct assignment for K_2CrO_4. Single crystal studies on scheelite structured compounds have been carried out by *Porto* and *Scott* (99, 195, 196) and others (97, 197, 198). The results of such a study for $CaMnO_4$ have been reproduced in Table 33 [see also Ref. (98)]. The Raman active phonons are classified for this case as follows:

$$\Gamma_t, \text{ Raman} = 3\,A_g + 5\,B_g + 5\,E_g$$

Here ν_4 is found at higher frequency than ν_2 as was also indicated in the 92/100Mo isotopic study (see Section 4.) which is, however, in contrast to the earlier interpretation of the powder data. Although the assignment of bands to irreducible representations is unequivocal, the same does not hold for differentiating translational modes from librational modes. This differentiation can be made by isotopic shift measurements as discussed in Section 4. In a recent investigation of $MgMoO_4$, C_{2h}^3 $Z = 8$, (199) complete agreement was found between the number of theoretically predicted and experimentally observed bands, 19 A_g and 17 B_g. Interestingly, ν_1 was found at about 50 cm^{-1} higher than for the free MoO_4^{2-} ion. The strong splitting for the ν_3 band has been attributed to a strong Mg—O interaction. The assignment of $MgMoO_4$, given in Table 34, shows that for the lower frequency bands it is impossible to differentiate internal and lattice mode vibrations. Some of the lowest frequency lattice bands are not given in this table.

A single crystal Raman spectrum of $NaReO_4$ recorded at low temperatures has recently been published by *Johnson et al.* (200). The measured phonons of this scheelite structure and their assignment are listed in Table 35.

6.2. Raman Band Intensities of Powdered Samples

The relative IR and Raman intensities can be very useful in making assignments. Usually, the $\nu_3(F_2)$ vibration appears as a strong broad band in the infra-red; whereas $\nu_1(A_1)$ appears weakly if at all for tetrahedral chalcogenoanions. In contrast, the ν_1 line is most intense in the Raman spectrum while ν_3 appears only weakly. For the bending region in the infra-red, ν_4 (F_2) is usually a strong to medium strong band, but $\nu_2(E)$, in general, is observed as a weak band if it is observed at all. Only recently the relative Raman intensities have been explained with respect to these two bands. Solution data and theoretical calculations have shown clearly that ν_2 must be more intense than ν_4 (44). This rule can also be extended to the Raman spectra of solid samples (121) under certain conditions (see for instance Na_2MoO_4).

The $\nu_3(F_{1u})$ infra-red band of hexaoxometallates always appears as a strong broad band. The most intense band in the Raman effect can be attributed to the totally symmetric $\nu_1(A_{1g})$ vibration.

6.3. Raman Band Intensities of Single Crystals

There are very few systematic studies of band intensities for single crystal Raman spectra. *Carter* and *Bricker* (79) have reported, for example, the size of the change of the polarizability tensor components, α_{ij}, for K_2CrO_4, which have been reproduced in Table 36. From this one can see that the A_g components of the ν_1 vibration shows an identical change for all three α_{jj} components, *i.e.* the amplitude of the motion of the oxygen atoms is expected to be practically the same in all three directions. One interesting result is the difference between the A_g phonons which belong to the ν_3 and ν_4 vibrations. The ν_3 component at higher frequency has its largest amplitude in the z-direction; whereas, for the other component, the largest amplitude is in the y-direction. For the ν_4 the higher frequency component has its largest amplitude in the z-direction, while the other ν_4 component has its largest amplitude in both the x- and the y-directions. The distinctions for the vibrational components with the same symmetry result most likely from the different environment in the different directions.

6.4. Band Widths in Spectra of Solid State Samples

Generally, infra-red bands are quite broad in the solid state. The following reasons are given as the explanation for this behavior of the bands (201—203):

I) The anharmonicity terms of the potential energy function,

II) correlation of the vibrational motions for the adjacent anions in the unit cell,

III) coupling of the internal with the external vibrations,

IV) overlapping of the absorption with the reflection process,

V) disorder in the crystal,

VI) other effects such as particle size.

It is remarkable, however, that the band width at half-height can be so markedly different for different anions. A qualitative reason has been put forward already to explain this behavior (203). The most important conclusion is that the anion-cation interactions within the unit cell grow with increasing formal charge of the anion, as in the series $AMnO_4$, A_2CrO_4, A_3VO_4. Furthermore, the electronic structure has also been shown to affect the bandwidth (203). Also, the more highly charged cations produce a clear splitting of the degenerate anions but on the other hand they lead to no wider band widths (203).

Raman spectra in general, do not show the same band-broading behavior.

6.5. Vibrational Spectra of Mixed Crystals

As was mentioned in the preceding section, the infra-red stretching vibration bands of chalcogenoanions appear usually as more or less smeared-out bands. In contrast, the infra-red spectra of small amounts of polyatomic anions isolated in alkali halide host lattices have very sharp bands [see results in Refs. (204—209)]. Thus, one can clearly see that the bands are better defined when the anion can be effectively isolated and thereby eliminate the coupling between neighboring identical ions as well as reducing the other effects discussed earlier. More will be said about this in the next section.

We would now like to discuss the preparation and spectroscopic investigation of mixed crystals from isostructural substances with different anions and identical cations. This technique was first developed by *Tarte* (49, 210—212) for the investigation of silicates and germanates. It was immediately obvious that it was a very useful tool and was rapidly picked up by other research groups. In the case of isostructural compounds with the same cations but different anions it is possible to substitute isomorphically one anion for another. Therefore, if the vibrations of the host and guest anions are separated enough in frequency, the internal vibrations of the guest can be measured without correlation field splitting.

Several oxoanions of transition metal elements have been spectroscopically investigated using this technique. *Steger* and *Danzer* (*213*) have studied mixed crystals of CrO_4^{2-} or MnO_4^- with sulphates, phosphates, arsenates, and selenates. However, the main point of this discussion is definitely the interpretation of the spectra of the oxoanions of the main group elements. $KMnO_4$ has also been measured mixed with $KClO_4$ (*214*). As expected, ν_3 and ν_4 were observed to split into three bands and ν_1 was clearly visible. The ν_2 components could, however, not be observed. In addition, the NH_4MnO_4-$BaSO_4$ mixed crystal has been studied by infra-red spectroscopy (*109*). Even at the concentrations used in this case it was still possible to see the correlation field effects.

Several interesting examples of the possibilities of this technique are provided by the systems $Ba_3(VO_4)_2/Ba_3(PO_4)_2$ and $Sr_3(VO_4)_2/Sr_3(PO_4)_2$ (*175*). Here, for the first time it was possible to determine clearly the position of the $\nu_1(A_1)$ vibration for Ba- and Sr-orthovanadate and to prove that the two bands observed for pure orthovanadates can be treated as the two predicted components of the $\nu_3(F_2)$ vibration. Thus, the mixed crystal studies also showed that the ν_1 band lies between the two ν_3 components and that it is not observed in the pure substance, not only because of its weaker intensity but also because it is overlapped by the very broad E component of the ν_3 vibration.

A further example is the investigation of the IR spectrum of normal II—III spinels (*177, 179*). In these oxide-systems, divalent as well as trivalent metal are interspersed. Thereby a clear and satisfactory assignment of the spectrum is possible for these complicated systems.

The measured absolute frequencies for mixed crystals are different from those measured for the pure substance because the isolated guest ion is contained in a lattice with different structural parameters from its own. A step by step frequency change is observable for a progressive series of compounds (*213*).

6.6. Vibrational Spectra of Anions Present in Low Concentration in Host Lattices

As already mentioned above, it is possible to include small quantities of a particular anion in the lattice of an alkalihalide. The alkalihalides, because of the low frequency of their absorptions, are particularly suited to be host lattices; however, salts such as sulphates, nitrates, perchlorates, to name a few, have also been used as host lattices.

Transition element chalcogenometallates have up to now been studied only infrequently. Recently, *Becher* (*201*) has begun a systematic investigation of these systems. To date, CrO_4^{2-}, MoO_4^{2-}, and WO_4^{2-} have been substituted into various isomorphous sulfate lattices and measured in the infra-red. He has also discussed general observation such as splitting of the bands

Müller et al. (*215*) have isolated $^{53}CrO_4^{2-}$, $^{50}CrO_4^{2-}$, $^{92}MoO_4^{2-}$, $^{100}MoO_4^{2-}$ in isomorphous alkali sulfates and measured the metal isotope shifts for $\nu_3(F_2)$. In another paper, the chromate(VI) ion was isolated in a KBr lattice with different divalent cations in conjunction (*216*). In this case it was quite obvious that ion pairs were isolated in the host lattice. This possibility must always be taken into consideration.

The advantage of this method lies in the fact that the bands are sharp and can be accurately measured.

6.7. Measurements at High Pressure

The vibrational spectra of inorganic molecular crystals of binary compounds of the type AB and AB_2, as well as ionic crystals of complex anions and cations, have been studied recently under pressures up to 70 Kbar (217–219). By this technique it is possible to differentiate between internal and lattice vibrations (220) since lattice modes have a greater dependence on pressure.

The Raman spectrum of $CaMoO_4$ and $CaWO_4$ (198) has been measured up to about 40 Kbar with a $d\nu/dp$ of between $0.1-1$ cm$^{-1} \cdot$ Kbar^{-1}. A new high pressure phase was discovered in the course of this study. Attempts to predict the sign of $d\nu/dp$ for the internal vibration by the Davydov-splitting model fell through (196).

Interestingly, $\nu_3(F_2)$ does not split when $KMnO_4$ is measured in the infra-red under pressure, while the deformation vibrations, $\nu_4(F_2)$, split into three bands (217). Under normal pressure only two bands are observed. The intensity of the bands at higher frequencies decreases with increasing pressure. The $\nu_1(A_1)$ band for $KMnO_4$ which was weak originally in the IR is observed to disappear when measured under pressure. $dI/dp < 0$ has frequently been observed for totally symmetric vibrations. In this way the difference between symmetric and anti-symmetric vibrations can be determined through the pressure dependence of intensity. This can be especially useful for assigning site and factor group vibrations measured for a powder.

6.8. Studies of Phase Transitions

Phase transitions can be studied by vibrational spectroscopy by means of changes of pressure or temperature. Other than the example of $CaMoO_4$ (198) given in Section 6.7. where a new phase was found at high pressure; little is known for chalcogenometallates along this line of study.

The temperature influence on phase transitions has been investigated for numerous ammonium salts of different anions. For $(NH_4)_2CrO_4$, three phases have been found and by the corresponding dichromate, two (221). These results are thought to be due to reorientation of the ammonium cations.

6.9 ATR and MIR Measurements

Solid state spectra can be measured by the Attenuated Total Reflectance or Multiple Internal Reflection methods as well as by simple transmission techniques (222). Brooker (223) has reported the discovery of the translational vibrations of $NaNO_2$, $NaNO_3$, and $CaCO_3$ (calcite). The orientation of the bonds in $(UO_2)(NO_3)_2 \cdot$ 6 H_2O has been ascertained from the spectra obtained in this manner (223).

The reflection spectrum of $DyVO_4$ in the far IR has also been reported (224) and the ATR spectrum of $BaCrO_4$ (225) measured.

6.10 Resonance Raman Spectra

The Resonance Raman Effect (RRE) ca be observed when a molecule is excited by light with a frequency which falls under an obsorption band of the molecule. Whereas an excitation of this type commonly produces fluorescence for the gas phase, the fluorescence is usually suppressed for solutions, pure liquids, and solid state samples. The Pre-Resonance Raman Effect (PRRE) is observed if the exciting line comes close to, but is not overlapping with an absorption band.

RR spectra have been recorded for MnO_4^- (226—228), CrO_4^{2-} (225, 228), MoS_4^{2-} (229, 230), VS_4^{3-} (231), WSe_4^{2-} (232), and $MoOS_4^{2-}$ (233). Additionally, the PRRE spectrum for CrO_4^{2-} has been recorded (228). The RRE has been observed using all eight of the visible Ar^+ laser lines, all of which lie on the short wavelength side of the first exciting line. Using the 4880 Å exciting line, seven $v_1(A_1)$ overtones with slightly decreasing intensity were recorded. The half-band width of the overtones increased with increasing vibrational quantum number. In the case of the 3638 Å UV laser line, nine $v_1(A_1)$ overtones were recorded for the RR spectrum of solid K_2CrO_4.

Using the visible Ar^+ laser lines, the RRE has been also observed for MoS_4^{2-} (229, 230). One overtone was recorded with the 5145 Å exciting line, two overtones with 4880 Å, three with 4727 Å, four with 4658 Å and five overtones were observed with 4579 Å. From the measured overtones of v_1 (A_1) for MnO_4^-, CrO_4^{2-}, MoS_4^{2-}, WSe_4^{2-} the anharmonicities, in cm^{-1}, can be determined as follows:

	x_{11}	$v_1(A_1)$	Ref.
MnO_4^- (solution)	1.0 ± 0.2	837.5	(228)
$KMnO_4$ (solid)	1.1 ± 0.2	842.5	(228)
K_2CrO_4 (solid)	0.71 ± 0.1	853	(228)
$BaCrO_4$ (solid)	5.5 ± 2.0 (?)	864	(225)
MoS_4^{2-} (solution)	0 ± 0.5	454	(230)
WSe_4^{2-} (solution)	0 ± 0.5	281	(232)

It would seem to us to be extremely interesting to investigate the influence of the cation on the anharmonicity constant of internal vibrations as seen, for example, for K_2CrO_4 and $BaCrO_4$ above. If there is no error in the measurements for chromate, then this influence found for Ba^{2+} is remarkably large. We hope to measure this effect systematically.

6.11. Influence of Sample Preparation and Measurement Temperature

Spectra of powder samples can be measured using different methods which have been described and discussed in detail elsewhere (234, 235). It needs only to be indicated here that the technique used has a decided influence on the resulting spectrum.

For the normal pellet technique where an alkali halide is used as pelleting material, reactions could occur between the matrix and the sample under study. Most of these reactions are found to be exchange reactions so that the solid compound is no longer seen in the spectrum; what is seen are individual ions isolated in the alkali halide lattice [see for example (236—239)]. Such exchange reactions have been observed for tetraoxoanions of the transition metals, as for example in chromates (69), perrhenates (119), and permanganates (108). Even "redox" processes in KBr are possible [35] as with $Ba(MnO_4)_2$ which is probably promoted by moist samples [see also (240)].

The particle size as well as the method and duration of grinding can have an influence on the spectrum. An example of these effects can be found in Ref. (241) and references contained therein. Considerable grinding is necessary to achieve reproducible data for Ba_2CoO_4 (41) and various divanadates (242). Other general information about the pellet technique is given in (243).

Improved spectra can be obtained through the cooling of the sample which often produces a more distinct splitting of degenerate bands. Bands which are weakly seen or seen only as shoulders at ambient temperature often become quite apparent at lower temperatures. However, chalcogenometallates of the transition metals have seldom been measured at low temperatures. A few examples are given in (244) and (245).

Acknowledgements. We wish to thank the Minister für Wissenschaft und Forschung des Landes Nordrhein-Westfalen, the Nato, the Deutsche Forschungsgemeinschaft, the Fonds der Chemischen Industrie, the Consejo Nacional de Investigaciones Cientificas y Tecnicas de la Republica Argentina for financial support. Two of the authors (*E. J. Baran* and *R. O. Carter*) wish to thank especially the Alexander von Humboldt-Stiftung for fellowships during their stay in the Bundesrepublik Deutschland. Thanks are also extended to Mrs. *Claudia M. de Baran*, Prof. Dr. *P. J. Aymonino*, Dr. *K. H. Schmidt* and Mrs. *Margaret B. Carter* for their generous help in the preparation of this survey.

7. Appendix

Correlations between the point group T_d and subgroups

Site groups	ν_1	ν_2	ν_3	ν_4
T_d	$A_1(R)$	$E(R)$	$F_2(R, IR)$	$F_2(R, IR)$
T	$A(R)$	$E(R)$	$F(R, IR)$	$F(R, IR)$
S_4	$A(R)$	$A(R) + B(R, IR)$	$B(R, IR) + E(R, IR)$	$B(R, IR) + E(R, IR)$
D_{2d}	$A_1(R)$	$A_1(R) + B_1(R)$	$B_2(R, IR) + E(R, IR)$	$B_2(R, IR) + E(R, IR)$
C_{3v}	$A_1(R, IR)$	$E(R, IR)$	$A_1(R, IR) + E(R, IR)$	$A_1(R, IR) + E(R, IR)$
C_3	$A(R, IR)$	$E(R, IR)$	$A(R, IR) + E(R, IR)$	$A(R, IR) + E(R, IR)$
C_{2v}	$A_1(R, IR)$	$A_1(R, IR) + A_2(R)$	$A_1 + B_1 + B_2(R, IR)$	$A_1 + B_1 + B_2(R, IR)$
D_2	$A(R)$	$2A(R)$	$B_1 + B_2 + B_3(R, IR)$	$B_1 + B_2 + B_3(R, IR)$
C_2	$A(R, IR)$	$2A(R, IR)$	$A + 2B(R, IR)$	$A + 2B(R, IR)$
C_s	$A'(R, IR)$	$A' + A''(R, IR)$	$2A' + 2A''(R, IR)$	$2A' + A''(R, IR)$
C_1	$A(R, IR)$	$2A(R, IR)$	$3A(R, IR)$	$3A(R, IR)$

Table 1. IR data (in cm^{-1}) for orthovanadates with divalent cations

	Ref.	ν_1	ν_2	ν_3	ν_4	Space group	Site group	Factor group	Stretching vibrations			
									$Z_{obs.}$		$Z_{theoret.}$	
									ν_1	ν_3	ν_1	ν_3
Mg$_2$(VO$_4$)$_3$	(51)	915	a)	860 831	a)	D$_{2h}^{18}$	C$_s$	D$_{2h}$	1	2	1	3
Ca$_3$(VO$_4$)$_2$	(51)	872	a)	841 810 750	a)	C$_{2h}^{6}$	b)	D$_{3d}$	1	3	b)	
Sr$_3$(VO$_4$)$_2$	(51, 52)	—	305	898 820	388 355	D$_{3d}^{5}$	C$_{3v}$	D$_{3d}$	—	2	1	2
Ba$_3$(VO$_4$)$_2$	(51, 52)	—	300	853 803	377 355	D$_{3d}^{5}$	C$_{3v}$	D$_{3d}$	—	2	1	2
Pb$_3$(VO$_4$)$_2$	(53)	815(?)	308 300	855 760	363 345	C$_{2h}^{6}$ (?)	C$_1$	C$_{2h}$	1	2	1	3

a) Measured only to 600 cm^{-1}.
b) See text.

107

A. Müller, E. J. Baran, and R. O. Carter

Table 3. IR data (in cm^{-1}) for orthovanadates of the lanthanides

	Ref.	ν_1	ν_2	ν_3	ν_4	Space group	Site group	Factor group	Stretching vibrations			
									$Z_{obs.}$		$Z_{theoret.}$	
									ν_1	ν_3	ν_1	ν_3
CeVO$_4$	(59)	—	—	860 808	442	D_{4h}^{19}	D_{2d}	D_{4h}	—	2	0	2
PrVO$_4$	(59)	—	—	870 808	441	D_{4h}^{19}	D_{2d}	D_{4h}	—	2	0	2
NdVO$_4$	(59)	—	—	865 802	443	D_{4h}^{19}	D_{2d}	D_{4h}	—	2	0	2
SmVO$_4$	(59)	—	—	870 811	445	D_{4h}^{19}	D_{2d}	D_{4h}	—	2	0	2
EuVO$_4$	(59)	—	—	865 807	443	D_{4h}^{19}	D_{2d}	D_{4h}	—	2	0	2
GdVO$_4$	(59)	—	—	870 810	448	D_{4h}^{19}	D_{2d}	D_{4h}	—	2	0	2
LaVO$_4$	(60)	875	a)	848 835 820 806 772	a)	C_{2h}^{5}	C_1	C_{2h}	1	5	1	3

a) The assignment of the bending region is unclear (see text).

Table 2. IR data (in cm^{-1}) for orthovanadates with monovalent cations

	Ref.	ν_1	ν_3
Li_3VO_4	(56)	?	884, 803
Ag_3VO_4	(57)	782	738, 718
Tl_3VO_4	(58)	840	790, 720

Table 4. IR data (in cm^{-1}) for orthovanadates with the apatite structure (62)

	ν_1	ν_3	ν_4
$Ca_5(VO_4)_3F$	850	882, 820	410, 390, 368
$Ca_5(VO_4)_3Cl$	850	872, 812	415, 363
$Ca_5(VO_4)_3Br$	848	870, 810	417, 365
$Sr_5(VO_4)_3F$	842	872, 812	397, 384, 358
$Sr_5(VO_4)_3Cl$	839	867, 813	403, 364
$Sr_5(VO_4)_3Br$	837	861, 808	405, 365
$Ba_5(VO_4)_3F$	830	850, 796	390, 380, 356
$Ba_5(VO_4)_3Cl$	828	848, 794	393, 378, 357
$Ba_5(VO_4)_3Br$	826	846, 786	392, 378, 360

Table 5. IR and Raman data (in cm^{-1}) for $YNbO_4$ and $YTaO_4$ (68)

	$YNbO_4$		$YTaO_4$	
	Raman	IR	Raman	IR
$\nu_1(A_1)$	832	800	825	810
$\nu_2(E)$	350	330	345	325
	340		320	
$\nu_3(F_2)$	715	720	720	720
	695	655	705	660
	675	590	670	605
	650	540	655	550
	560			
$\nu_4(F_2)$	480	470	480	470
	455	400	450	415
	435	385	375	390
	385	360		380
				360

Table 6. Vibrational data for chromates(VI)

	Ref.	ν_1	ν_2	ν_3	ν_4	Space group	Site group	Factor group	$Z_{obs.}$ ν_1	$Z_{obs.}$ ν_3	$Z_{theoret.}$ ν_1	$Z_{theoret.}$ ν_3
Na$_2$CrO$_4$[a]	(80)	850	350 361	899 932	390 408	D_{2h}^{17}	C_{2v}	D_{2h}	1	2	1	3
K$_2$CrO$_4$[a]	(78)	855	351	871 883 908	395	D_{2h}^{16}	C_s	D_{2h}	1	3	1	3
Rb$_2$CrO$_4$[a]	(78)	846	347	868 876 895	387	D_{2h}^{16}	C_s	D_{2h}	1	3	1	3
Cs$_2$CrO$_4$[a]	(78)	840	344	865 886	385	D_{2h}^{16}	C_s	D_{2h}	1	2	1	3
Ag$_2$CrO$_4$[a]	(82a)	812	355 339	876 856 849	375	D_{2h}^{16}	C_s	D_{2h}	1	3	1	3
Tl$_2$CrO$_4$[b]	(71)	821	[c)	867 843	[c)	D_{2h}^{16}	C_s	D_{2h}	1	2	1	3

110

Table 6 (continued)

	Ref.	ν_1	ν_2	ν_3	ν_4	Space group	Site group	Factor group	Stretching vibrations			
									$Z_{obs.}$		$Z_{theoret.}$	
									ν_1	ν_3	ν_1	ν_3
$(NH_4)_2CrO_4$ [a]	(80)	847	345	879	382 410	C_{2h}^3	C_s	C_{2h}	1	1	1	3
$CaCrO_4$ [b]	(71)	—	c)	925 892	c)	D_{2h}^{19}	D_{2d}	D_{2h}	—	2	0	2
$SrCrO_4$ [b]	(71)	854 844	c)	925 911 887 874	c)	C_{2h}^5	C_1	C_{2h}	2	4	1	3
$BaCrO_4$ [b]	(70)	860	(?)	949 894 873	419 389 375	D_{2h}^{16}	C_s	D_{2h}	1	3	1	3
$PbCrO_4$ [b]	(71)	832	c)	907 864 853	c)	C_{2h}^5	C_1	C_{2h}	1	3	1	3

a) Raman.
b) IR.
c) Measured in the NaCl-region only.

Table 7. Raman data (in cm^{-1}) for Na$_2$MoO$_4$ and Na$_2$WO$_4$ (84)

	Na$_2$MoO$_4$	Na$_2$WO$_4$
$\nu_1(A_1)$	890	928
$\nu_2(E)$	303	312
$\nu_3(F_2)$	808	811
$\nu_4(F_2)$	381	375

Table 8. IR and Raman data (in cm^{-1}) of some monoclinic molybdates and tungstates (84)

	ν_1	ν_3	ν_4	ν_2	
K$_2$MoO$_4$	888	853	371	317	R
		823	347	338	
	885	855	332	311	IR
		820			
Rb$_2$MoO$_4$	883	841	362	318	R
		820	342	313	
		813	334		
	890	828	340	302	IR
				318	
K$_2$WO$_4$	924	850	358	326	R
		823	320		
	921	822	336	295	IR
		855	316		
Rb$_2$WO$_4$	922	840	350	325	R
		823		316	
		818			
	918	820	330	295	IR
			310	316	
			295		

Table 9. IR and Raman data (in cm^{-1}) for orthorhombic molybdates and tungstates (84)

	ν_1	ν_3	ν_4	ν_2	
Rb$_2$MoO$_4$	889	848	339	315	R
		825	328		
		816			
	887	844	340	303	IR
		830	321		
Cs$_2$MoO$_4$	882	835	334	310	R
		820	327	306	
		813	321		
	881	825	334	315	IR
			310	305	
Cs$_2$WO$_4$	921	835	331	319	R
		820	315		
		815	306		
	921	824	324	297	IR
			313		
			297		

Table 10. IR data (in cm^{-1}) for permanganates with monovalent cations

	Ref.	ν_1	ν_2	ν_3	ν_4	Space group	Site group	Factor group	Stretching vibrations				
									$Z_{obs.}$			$Z_{theoret.}$	
									ν_1	ν_3		ν_1	ν_3
KMnO$_4$	(107)	845	?	930 907	400 385 381	D_{2h}^{16}	C_s	D_{2h}	1	2		1	3
RbMnO$_4$	(107)	845	?	920 910 900	400 387 382	D_{2h}^{16}	C_s	D_{2h}	1	3		1	3
CsMnO$_4$	(107)	845	?	920 910 900	400 390 383	D_{2h}^{16}	C_s	D_{2h}	1	3		1	3
NH$_4$MnO$_4$	(107, 109)	841	?	923 911 896	400 387 382	D_{2h}^{16}	C_s	D_{2h}	1	3		1	3
LiMnO$_4 \cdot 3$ H$_2$O	(111)	845	a)	920 895	a)	C_{6v}^4	C_{3v}	C_{6v}	1	2		1	2
AgMnO$_4$	(107)	800	a)	898 882 852	a)	C_{2h}^5	C_1	C_{2h}	1	3		1	3
[As(C$_6$H$_5$)$_4$]MnO$_4$	(110)	—	?	899	?	S_4^2	S_4	S_4	—	1		0	2
[P(C$_6$H$_5$)$_4$]MnO$_4$	(110)	—	?	901	?	S_4^2	S_4	S_4	—	1		0	2

a) Only measured to 400 cm^{-1}.

Table 11. IR data (in cm⁻¹) for permanganates with divalent cations

	Ref.	ν_1	ν_2	ν_3	ν_4	Space group	Site group	Factor group	Stretching vibrations			
									$Z_{obs.}$		$Z_{theoret.}$	
									ν_1	ν_3	ν_1	ν_3
Ba(MnO₄)₂	(107)	843	350	940 930 920 880	398 378 364	D_{2h}^{24}	C_2	D_{2h}	1	4	1	3
Sr(MnO₄)₂ · 3 H₂O	(111)	845	a)	945 930 920 900 890	a)	T^4	C_3	T	1	5	1	2
Mg(MnO₄)₂ · 6 H₂O	(108)	839	b)	904	b)	C_{2v}^7	C_1	C_{2v}	1	1	1	3
Zn(MnO₄)₂ · 6 H₂O	(108)	839	b)	900	b)	C_{2v}^7	C_1	C_{2v}	1	1	1	3
Ni(MnO₄)₂ · 6 H₂O	(108)	837	b)	904	b)	C_{2v}^7	C_1	C_{2v}	1	1	1	3
Cd(MnO₄)₂ · 6 H₂O	(112)	839	b)	902	b)	C_{2v}^7	C_1	C_{2v}	1	1	1	3
Cu(MnO₄)₂ · 2 H₂O	(108)	838	b)	906	b)	unknown			—	—	—	—

a) Measured only to 400 cm⁻¹.
b) Measured only to 600 cm⁻¹.

A. Müller, E. J. Baran, and R. O. Carter

Table 12. IR data (in cm^{-1}) for permanganates with complex cations

	Ref.	ν_1	ν_2	ν_3	ν_4
$[Co(NH_3)_6](MnO_4)_3$	(75, 113)	?	?	913 897	400
$[Cr(NH_3)_6](MnO_4)_3$	(75)	?	?	923 914 892	400
$[Ni(NH_3)_6](MnO_4)_2$	(75)	841	a)	910 889	a)
$[Cu(NH_3)_4](MnO_4)_2$	(76)	836	?	918 891	402
$[Cd(NH_3)_4](MnO_4)_2$	(76)	?	a)	918 907	a)
$[Zn(NH_3)_4](MnO_4)_2$	(76)	?	?	920 910 898	385

a) Measured only to 400 cm^{-1}.

Table 13. IR data for pertechnetates

	Ref.	ν_1	ν_2	ν_3	ν_4	Space group	Site group	Factor group	Stretching vibrations			
									$Z_{obs.}$		$Z_{theoret.}$	
									ν_1	ν_3	ν_1	ν_3
NH_4TcO_4	(107)	—	348	925 900	329 317	C_{4h}^6	S_4	C_{4h}	—	2	0	2
$KTcO_4$	(115)	—	351	910	328 313	C_{4h}^6	S_4	C_{4h}	—	1	0	2
$CsTcO_4$	(107)	?	347	920 910 895	330 325	D_{2h}^{16}	C_s	D_{2h}	—	3	1	3
$TlTcO_4$	(107)	?	347	900 875 860	332 327 322	D_{2h}^{16}	C_s	D_{2h}	—	3	1	3
$AgTcO_4$	(107)	—	a)	900 865	a)	C_{4h}^6	S_4	C_{4h}	—	2	0	2

a) Measured only to 400 cm^{-1}.

117

Table 14. IR data (in cm^{-1}) for perrhenates of known structure

	Ref.	ν_1	ν_2	ν_3	ν_4	Space group	Site group	Factor group	Stretching vibrations			
									$Z_{obs.}$		$Z_{theoret.}$	
									ν_1	ν_3	ν_1	ν_3
KReO$_4$	(107)	—	317 304	929 915 898	361	C_{4h}^6	S_4	C_{4h}	—	3	0	2
NH$_4$ReO$_4$	(107)	—	318 303	930 914 904	361	C_{4h}^6	S_4	C_{4h}	—	2	0	2
RbReO$_4$	(107)	—	316 308	935 915	350	C_{4h}^6	S_4	C_{4h}	—	2	0	2
CsReO$_4$	(107)	?	320 315	935 920 907	335	D_{2h}^{16}	C_8	D_{2h}	—	3	1	3
AgReO$_4$	(107)	—	296 278	910 880	315?	C_{4h}^6	S_4	C_{4h}	—	2	0	2
NaReO$_4$	(84)	—	385	925 901	301 289	C_{4h}^6	S_4	C_{4h}	—	2	0	2
TlReO$_4$	(107)	965?	307	920 900 890	347	D_{2h}^{16}	C_8	D_{2h}	1	3	1	3

Table 15. Raman data (in cm^{-1}) for perrhenates of unknown structure

	Ref.	ν_1	ν_3	$\nu_2 + \nu_4$
LiReO$_4$	(120)	994[a])	890[a])	321
			910	347
			938	358
			949	
			956	
			966	
Mg(ReO$_4$)$_2$	(120)	1024	978	343
			954	380
			911	
Zn(ReO$_4$)$_2$	(120)	1025	954	339
			891	323
Ba(ReO$_4$)$_2$	(120)	978	925	364
			901	346
			896	339
				332
				327
Cd(ReO$_4$)$_2$	(120)	974	939	369
			910	339
			894	325
			882	
			859	
Ca(ReO$_4$)$_2$	(120)	991	969	352
			940	344
			914	338
			906	314
Sr(ReO$_4$)$_2$	(120)	997	960	354
		986	932	343
			914	339
			904	335
				315
Pb(ReO$_4$)$_2$	(120)	983	961	342
		972	923	325
			912	312
			885	
			868	
			850	

[a]) Values taken from the IR.

Table 16. Raman data (in cm^{-1}) for crystalline OsO_4 and RuO_4 [a]

	OsO_4 (123)	RuO_4 (123)
ν_1	961	878
ν_3	967,5	921
	951,5	905
	940,0	903
ν_2	342	333
ν_4	327	328
	319	321,5
	316	

[a] Compare with Refs. (84) and (121) in deformation vibration region.

Table 17. IR data (in cm^{-1}) for chromates(V)

	Ref.	ν_1	ν_2	ν_3	ν_4	Space group	Site group	Factor group	Stretching vibrations				
									$Z_{obs.}$		$Z_{theoret.}$		
									ν_1	ν_3	ν_1	ν_3	
$Ca_3(CrO_4)_2$	(40, 135)	863	a)	817 766 713	a)	D_{3d}^6?	b)	D_{3d}	1	3	b)		
$Sr_3(CrO_4)_2$	(135)	—	a)	839 783	a)	D_{3d}^5	C_{3v}	D_{3d}	—	2	1	2	
$Ba_3(CrO_4)_2$	(135)	—	a)	843 770	a)	D_{3d}^5	C_{3v}	D_{3d}	—	2	1	2	

a) Measured only to 600 cm^{-1}.
b) See text.

Table 18. IR data (in cm^{-1}) of alkalimanganates(VI) (138) from the NaCl spectral region

	ν_1	ν_3	Space group	Site group	Factor group	Stretching vibrations			
						$Z_{obs.}$		$Z_{theoret.}$	
						ν_1	ν_3	ν_1	ν_3
K_2MnO_4	811	856 838	D_{2h}^{16}	C_s	D_{2h}	1	2	1	3
Rb_2MnO_4	807	837	D_{2h}^{16}	C_s	D_{2h}	1	1	1	3
Cs_2MnO_4	802	830	D_{2h}^{16}	C_s	D_{2h}	1	1	1	3

Table 19. IR data (in cm^{-1}) for some ferrates(VI)

	Ref.	ν_1	ν_2	ν_3	ν_4	Space group	Site group	Factor group	Stretching vibrations			
									$Z_{obs.}$		$Z_{theoret.}$	
									ν_1	ν_3	ν_1	ν_3
K_2FeO_4	(70)	782	?	825 809	340 322	D_{2h}^{16}	C_s	D_{2h}	1	2	1	3
$BaFeO_4$	(70)	790	?	870 818 783	362 340 306	D_{2h}^{16}	C_s	D_{2h}	1	3	1	3
Cs_2FeO_4	(142)	771	?	800	332 322 310	D_{2h}^{16}	C_s	D_{2h}	1	1	1	3

Table 20. Vibrational data (in cm^{-1}) of some hexaoxometallates (when not marked the frequencies come from the IR spectra)

	$\nu_1(A_{1g})$	$\nu_2(E_g)$	$\nu_3(F_{1u})$	$\nu_4(F_{1u})$	$\nu_5(F_{2g})$	$\nu_6(F_{2u})$	Ref.
Li$_8$ZrO$_6$	—	—	520	430	—	—	(150)
Li$_8$HfO$_6$	—	—	530	430	—	—	(150)
Li$_7$NbO$_6$	—	—	610	420	—	—	(150)
Li$_7$TaO$_6$	—	—	610	410	—	—	(150)
Ba$_2$CaMoO$_6$	812(R)	650(R)	598	357	416(R)	—	(153)
Ba$_2$CaWO$_6$	832(R)	675(R)	628	327	410(R)	—	(153)
Li$_6$WO$_6$	(740)$^{a)}$	(450)$^{a)}$	620	425	(360)$^{a)}$	—	(150, 152)
α-Li$_6$ReO$_6$	(680)$^{a)}$	(505)$^{a)}$	620	425	(360)$^{a)}$	—	(152)
Li$_5$ReO$_6$	—	—	650	450	—	—	(149, 150)
Ca$_5$(ReO$_6$)$_2$	(712)$^{b)}$	(595)$^{b)}$	655	375	—	—	(149)
Sr$_5$(ReO$_6$)$_2$	(685)$^{b)}$	(565)$^{b)}$	628	365	(335)$^{b)}$	—	(149)
Sr$_4$Ca(ReO$_6$)$_2$	(685)$^{b)}$	(585)$^{b)}$	628	365	(340)$^{b)}$	—	(149)
Li$_8$PtO$_6$	—	(535)$^{a)}$	575$^{c)}$	475	(410)$^{a)}$	—	(149)
Mg$_3$Li$_2$PtO$_6$	—	(535)$^{b)}$	585$^{c)}$	450	—	—	(149)
Ca$_4$PtO$_6$	—	(530)$^{b)}$	575$^{c)}$	425	(345)$^{b)}$	(318)$^{b)}$	(149)
Sr$_4$PtO$_6$	—	(475)$^{b)}$	520$^{c)}$	410	(342)$^{b)}$	(315)$^{b)}$	(149)
Ba$_4$PtO$_6$	—	(434)$^{b)}$	470$^{c)}$	371	(321)$^{b)}$	(300)$^{b)}$	(149)

a) These vibration frequencies are forbidden for the site symmetry of the MO$_6$ unit.
b) Assignment uncertain.
c) A frequency difference of more than 100 cm^{-1} is unrealistic.

Table 21. Vibrational spectral data (in cm^{-1}) for tetrathiomolybdates and -tungstates (*156*)

	$\nu_1(A_1)$ $\nu_s(M—S)$	$\nu_3(F_2)$ $\nu_{as}(M—S)$	$\nu_2(E)$ $\delta_s(M—S)$	$\nu_4(F_2)$ $\delta_{as}(M—S)$	
$(NH_4)_2MoS_4$	455	477 469	180	194	R
	455	476	—	~215 195	IR
K_2MoS_4	456	482 475 469	180	198	R
	456	476	—	203 188	IR
Rb_2MoS_4	456	480 473 468	176	195	R
	456	476	—	199 188	IR
Cs_2MoS_4	456	479 470	176	190	R
	456	474	—	192 184	IR
$(NH_4)_2WS_4$	485	472 459	178	~190	R
	485	458	—	~215 185	IR
K_2WS_4	487	472 461 455	180	196	R
	487	461	—	200 181	IR
Rb_2WS_4	485	471 461 455	180	191	R
	485	461	—	194 180	IR
Cs_2WS_4	482	469 460 455	179	186	R
	482	471 459	—	188 178	IR

Table 22. Raman and IR data (in cm^{-1}) for some tetrathio- and tetraselenometallates

		$\nu_1(A_1)$	$\nu_2(E)$	$\nu_3(F_2)$	$\nu_4(F_2)$	Ref.
Tl_3VS_4	R	375	—	—	166,5	
	IR —		—	460	190	(161)
Tl_3NbS_4	R	408	—	—	157	
	IR —		—	421	168,5	(161)
Tl_3TaS_4	R	424	—	398	165	
	IR —		—	400	174	(161)
Tl_3VSe_4	R —		—	—	—	
	IR —		—	365	121	(161)
Tl_3NbSe_4	R	239	—	—	~ 92	
	IR —		—	316	107	(161)
Tl_3TaSe_4	R	249	—	274,5	~ 98	
	IR —		—	278	108,5	(161)
Rb_2MoSe_4	R	237	—	—	—	
	IR —		—	335	113	(161
Cs_2WSe_4	R	280,5	—	309	107	
	IR —		—	309	—	(161)
$[(C_6H_5)_4P]ReS_4$	R	501	—	—	—	
	IR —		200	488	200	(162)

Table 23. Vibrational data (in cm^{-1}) of compounds of the type $A_2MS_xSe_{4-x}$[a]) (165)

		$\nu_1(A_1)$	$\nu_2(A_1)$	$\nu_3(A_1)$	$\nu_4(E)$	$\nu_5(E)$	$\nu_6(E)$
Cs_2MoS_3Se[a])	IR	349	458	160	473	150	183
Cs_2MoSSe_3[a])	IR	471	—	121	342	—	121
Cs_2WSSe_3[a])	IR, R	468	282	108	311	150	108

[a]) See text concerning stoichiometry of the mixed crystals.

Table 24. Vibrational data (in cm^{-1}) for trithiomolybdates and trithiotungstates (*156*)

	$\nu_1(A_1)$ $\nu(MO)$	$\nu_2(A_1)$ $\nu_s(MS)$	$\nu_4(E)$ $\nu_{as}(MS)$	$\nu_5(E)$ $\varrho(MS_3)$	$\nu_3(A_1) + \nu_6(E)$ $\delta(MS_3)$		
K$_2$WOS$_3$	866	478	466 456	262 257	192	178	R
	868 846(?)		460	264 258	197	186	IR
Rb$_2$WOS$_3$	869	477	465 458	264 253	193	181	R
	869 845(?)	476	461	263 254	195	176 188	IR
Cs$_2$WOS$_3$	867	473	459	263 250	187	176	R
	872 845(?)	473	457	263 252	192	175 186	IR
K$_2$MoOS$_3$	851	468	498 485	263 253	194	181	R
	858		490 474	263 259		200	IR
Rb$_2$MoOS$_3$	852	467	482 473	263 253	194	181	R
	856	468(?)	474	264 255		198	IR
Cs$_2$MoOS$_3$	851	464	479 473	263 252	189	178	R
	854	465	473	263		197	IR

Table 25. Vibrational data (in cm^{-1}) for Cs$_2$MoOSe$_3$ and Cs$_2$WOSe$_3$ (*170*)

	Cs$_2$MoOSe$_3$		Cs$_2$WOSe$_3$	
	R	IR	R	IR
$\nu_1(A_1) = \nu(Me\!-\!O)$	—	858	—	878
$\nu_2(A_1) = \nu_s(Me\!-\!Se)$	293,5	291	290	294
$\nu_3(A_1) = \delta_s(MeSe_3)$	—	120	—	—
$\nu_4(E) = \nu_{as}(Me\!-\!Se)$	357	352,5	313	311
$\nu_5(E) = \varrho(MeSe_3)$	187	188	195	193
$\nu_6(E) = \delta_{as}(MeSe_3)$	—	120	—	—

Table 26. Vibrational data (in cm⁻¹) for chalcogenometallates with the form $MO_2X_2^{2-}$ (C_{2v})

	Tl₂MoO₂S₂ (158)		Tl₂WO₂S₂ (158)		(NH₄)₂MoO₂Se₂ (170)		(NH₄)₂WO₂Se₂ (170)		Cs₂WO₂Se₂ (170)	
	IR	R	IR	R	IR	R	IR	R	IR	R
ν_s(Me—O) (A₁)	844	845	880 865	875 853	836	—	863	849	891	885
ν_s(Me—X) (A₁)	~453	461	459	465	359	346	—	335	326	—
δ(MeO₂) (A₁)	291	296	293	304	280	285	283	282	277	276
δ(MeX₂) (A₁)	a)	179 190	a)	175 185	114	—	113	119	121	—
τ(A₂)	a)	256	a)	256	—	—	—	—	—	—
ν_{as}(Me—O) (B₁)	825	829	831	824	799	—	808	813	845	—
δ(MeOX) (B₁)	a)	235	a)	239	—	251	—	235	—	237
ν_{as}(Me—X) (B₂)	468	444 b)	445	455 434	342	336	323	315	313	323
δ(MeOX) (B₂)	a)	256	a)	256	—	—	—	156	—	—

a) Not measured.
b) Probable assignment.

Table 27. Raman and IR data (in cm⁻¹) for the MX_4 tetrahedra of compounds of the form Cu_3MX_4 (173)

		$\nu_s(M-X)$	$\nu_{as}(M-X)$	$\delta(MX_4)$
Cu_3VS_4	R	377	—	—
	IR	—	437	202
Cu_3NbS_4	R	406	—	199
	IR	—	430	—
Cu_3TaS_4	R	415	—	200
	IR	—	410	—
Cu_3VSe_4	R	—	—	—
	IR	—	341	133
Cu_3NbSe_4	R	239	—	—
	IR	—	315	—
Cu_3TaSe_4	R	245	—	122
	IR	—	282	—

Table 28. Cu—X stretching frequencies (in cm⁻¹) for compounds of the type Cu_3MX_4 (173)

		$\nu_s(CuX)$	$\nu_{as}(CuX)$
Cu_3VS_4	R	—	—
	IR	—	285
Cu_3NbS_4	R	267	—
	IR	—	244
Cu_3TaS_4	R	267	—
	IR	—	244
Cu_3NbSe_4	R	169	146, 135
	IR	—	142, 136
Cu_3TaSe_4	R	170	141
	IR	—	141

Table 29. IR frequencies (in cm⁻¹) and assignment of $^{40}CaMoO_4$ and $^{44}CaMoO_4$

$^{40}CaMoO_4$	$^{44}CaMoO_4$	$\Delta\nu$	Assignment
432	432	0	ν_2
329	326	—3	Mainly ν_4 but coupled with a translation vibration
284	282	—2	Mainly ν_4 but coupled with a translation vibration
237	224	—13	Translation
200	192	—8	Translation
153	153	0	Libration

A. Müller, E. J. Baran, and R. O. Carter

Table 30. Raman frequencies (in cm^{-1}) for CaMoO$_4$ (98)

^{40}CaMoO$_4$	^{44}CaMoO$_4$	$\Delta\nu$	Ca^{92}MoO$_4$	Ca^{100}MoO$_4$	$\Delta\nu$	
879	879	0	879	879	0	$\nu_1(Ag)$
848	848	0	852	844,5	—7,5	$\nu_3(Bg)$
794,5	794	—0,5	797,5	791	—6,5	$\nu_3(Eg)$
403,5	403	—0,5	403	401	—2	$\nu_4(Eg)$
392	392,5	0,5	393	390,5	—2,5	$\nu_4(Bg)$
323,5	323,5	0	323	323	0	$\nu_2(Ag)/\nu_2(Bg)$
~269	~268	—1	~267	267	0	$R(Eg)$
205	205	0	205	205	0	$R(Ag)$
191,5	183,5	—8	190,5	190,5	0	$T(Eg)$Ca/Ca
144	143	—1	145,5	142	—3,5	$T(Eg)$Mo/Mo
112	111,5	—0,5	113,5	109,5	—4	$T(Bg)$Mo/Mo

Table 31. Use of the product-rule for the "F_2vibration" of A$_2$MO$_4$ type compounds

		Numerical values	
		Observed ratios	Calculated ratios
Na$_2$MoO$_4$	IR	1.032	
Na$_2$MoO$_4$	R	1.033	1.034
Ag$_2$MoO$_4$	IR	1.034	
Ag$_2$MoO$_4$	R	1.036	
Ni$_2$GeO$_4$	IR	1.040	1.064
Co$_2$GeO$_4$	R	1.058	

Table 32. Single crystal Raman measurements (in cm^{-1}) for K$_2$CrO$_4$ (*194*) and (NH$_4$)$_2$WS$_4$ (*194 a*)

values (Solution CrO$_4^{2-}$)	Point group	Site symmetry	Factor group	K$_2$CrO$_4$	(NH$_4$)$_2$WS$_4$
	T_d	C_s	D_{2h}		
884	F_2	$2\,A' + A''$	B_{2g}	919	470
			Ag	905	—
			B_{2g}	886	458
			A_g	868	456
			B_{1g}	881	458
			B_{3g}	877	458
847	A_1	A'	$A_g + B_{2g}$	852	483
368	F_2	$2\,A' + A''$	B_{2g}	398	189
			A_g	397	189
			B_{2g}	388	177
			A_g	387	189
			B_{3g}	393	187
			B_{1g}	388	189
348	E	$A' + A''$	B_{2g}	352	177
			A_g	347	177
			B_{3g}	351	177
			B_{1g}	348	177

Table 33. Single crystal Raman measurements (in cm^{-1}) for CaMoO$_4$ (*197*)

$\nu_1\,A_{1g}$	878
$\nu_3\,B_g$	844
$\nu_3\,E_g$	797
$\nu_2\,A_g$	333
$\nu_2\,B_g$	339
$\nu_4\,B_g$	393
$\nu_4\,E_g$	401
$R(A_g)$	205
$R(E_g)$	263
$T(B_g)$	219
$T(B_g)$	110
$T(E_g)$	189
$T(E_g)$	145

A. Müller, E. J. Baran, and R. O. Carter

Table 34. Single crystal Raman measurements (in cm⁻¹) for MgMoO₄ (199)

	A_g	B_g
ν_1	970	
	959	
ν_3	906	912
	856	874
	754	774
ν_2, ν_4 and lattice vibrations	427	403
	385	371
	371	351
	345	322
	339	308
	332	290
	324	280
	308	
	275	

Table 35. Single crystal Raman measurements (in cm⁻¹) for NaReO₄ (200)

		77 K	294 K
ν_1	A_g	952	958
	B_u	—	—
ν_2	A_g	334	335
	B_g	320	325
	A_u	—	—
	B_u	—	—
ν_3	B_g	925	924
	E_g	888	888
	A_u	—	—
	E_u	—	—
ν_4	B_g	374	—
	E_g	378	372
	A_u	—	—
	E_u	—	—

Table 36. Relative magnitude of the changes in the α_{ij}-components of the Raman tensor for K_2CrO_4 (79)

	Frequency	Relative Magnitude[a]
ν_3	903 (A_g)	$zz > yy > xx$
	867 (A_g)	$yy > zz > xx$
ν_1	851 (A_g)	$xx \approx yy \approx zz$
ν_4	396 (A_g)	$zz > xx \approx yy$
	386 (A_g)	$xx \approx yy > zz$
ν_2	345 (A_g)	$xx \approx yy > zz$

[a] Qualitative estimation from the intensities of the bands for different orientations.

8. References

1. *Siebert, H.:* Anwendungen der Schwingungsspektroskopie in der Anorganischen Chemie. Berlin-Heidelberg-New York: Springer 1966.
2. *Lecomte, J.:* In: Handbuch der Physik (*S. Flügge,* ed.), Bd. 26. Berlin-Göttingen-Heidelberg: Springer 1958.
3. *Ross, S. D.:* Inorganic infrared and raman spectra. London: McGraw Hill 1972.
4. *Krasnov, K. S., Timoshinin, V. S., Danilova, T. G., Khandozhko, S. V.:* Molekulyarne postoyannie neoyanicheskikh molekul (ed. Khimiya). Leningrad: 1968.
5. *Nakamoto, K.:* Infrared spectra of inorganic and coordination compounds, 2nd. edit. New York: J. Wiley 1970.
6. *Craig, D. P., Walmsley, S. H.:* Excitons in molecular crystals: Theory and applications. London-New York: Benjamin 1968.
7. *Schutte, C. J. H.:* Fortschr. Chem. Forsch. *36,* 57 (1973).
8. *Hodgson, J. N.:* Optical absorption and dispersion in solids. London: Chapman and Hall 1970.
8a. *Sherwood, P. M. A.:* Vibrational spectroscopy of solids. Cambridge: Cambridge University Press 1972.
9. *Donovan, B., Angress, F.:* Lattice vibrations. London: Chapman and Hall 1971.
10. *Turrell, G.:* Infrared and Raman spectra of crystals. New York: Academic Press 1972.
11. *Fateley, W. G., Dollish, F. R., McDevitt, N. T., Bentley, F. F.:* Infrared and raman selection rules for molecular and lattice vibrations: The correlation method. New York: J. Wiley 1972.
12. *Lazarev, A. N.:* Vibrational spectra and structure of silicates (english translation). New York: Consultants Bureau 1972.
13. *Abelès, F.:* Optical properties of solids. Amsterdam: North Holland Publish. Co. 1972.
14. *Bradley, C. J., Cracknell, A. P.:* Mathematical theory of symmetry in solids: Representation theory for point groups and space groups. Oxford: Clarendon Press 1972.
15. *Becher, H. J.:* Angew. Chem. Intern. Ed. Engl. *11* 26 (1972).
16. *Mitra, S. S., Gielisse, P. J.:* Progress in infrared spectroscopy, Vol. 2, p. 47 (ed. *Szymansky*). New York: Plenum Press 1963.
17. *Loudon, R.:* Advan. Phys. *13,* 423 (1964).
18. *Matossi, F.:* Gruppentheorie der Eigenschwingungen von Punktsystemen. Berlin-Göttingen-Heidelberg: Springer 1961.
19. *Halford, R. S.:* J. Chem. Phys. *14,* 8 (1946).
20. *Fateley, W. G., McDevitt, N. T., Bentley, F. F.:* Appl. Spectry. *25,* 155 (1971).
21. *Gilson, T. R., Hendra, P. J.:* Laser Raman spectroscopy. London: Wiley 1970.
22. *Bhagavantam, S., Venkatarayudu, T.:* Theory of groups and its application to physical problems. Bangalore: Bangalore Press 1951; New York: Academic Press 1969; Proc. Roy. Soc. *130A,* 259 (1931); Proc. Indian Acad. Sci. *9A,* 224 (1939).
23. *Bertie, J. E., Bell, J. W.:* J. Chem. Phys. *54,* 160 (1970).
24. *Bertie, J. E., Kopelman, R.:* J. Chem. Phys. *55,* 3613 (1971).
25. *Davydov, A. S.:* Theory of molecular excitons. New York: Plenum Press 1971.
26. *Irish, D. E., Brooker, M. H.:* Appl. Spectry. *27,* 395 (1973).
27. *Fateley, W. G.:* Appl. Spectry. *27,* 395 (1973).
28. *Adams, D. M., Newton, D. C.:* J. Chem. Soc. *A1970,* 2822.
29. *Adams, D. M., Newton, D. C.:* Tables for factor group analysis. London: Beckman—RIIC. Ltd. 1970.
30. *Adams, D. M.:* Coord. Chem. Rev. *10,* 183 (1973).
31. *Damen, T. C., Porto, S. P. S., Tell, B.:* Phys. Rev. *142,* 570 (1966).
32. *Griffith, W. P.:* Coord. Chem. Rev. *5,* 459 (1970).
33. *Müller, A., Diemann, E.:* MTP Int. Rev. Sci., Inorg. Series 2, Vol. 5.
34. *Gonzalez-Vilchez, F., Griffith, W. P.:* J. Chem. Soc. Dalton *1972,* 1416.
35. *Baran, E. J., Aymonino, P. J.:* Spectrochim. Acta *24A,* 291 (1968).
36. *Krebs, B., Müller, A., Roesky, H. W.:* Mol. Phys. *12,* 469 (1967).
37. *Müller, A., Krebs, B.:* Spectrochim. Acta *23A,* 1591 (1967).
38. *Krebs, B., Müller, A.:* Z. Chem. *7,* 243 (1967).

39. *Müller, A., Baran, E. J., Aymonino, P. J.:* Anales Asoc. Quim. Arg. *56*, 85 (1968).
40. *Baran, E. J., Aymonino, P. J.:* Anales Asoc. Quim. Arg. *56*, 91 (1968).
41. *Baran, E. J.:* Z. Anorg. Allgem. Chem. *399*, 57 (1973).
42. *Müller, A., Diemann, E.:* Chem. Ber. *102*, 945 (1969).
43. *Baran, E. J.:* Z. Naturforsch. *27a*, 1000 (1972).
44. *Weinstock, N., Schulze, H., Müller, A.:* J. Chem. Phys. *59*, 5063 (1973).
45. *Dubey, B. L., West, A. R.:* Nature *235*, 155 (1972).
46. *Bland, J. A.:* Acta Cryst. *14*, 875 (1961).
47. *Tarte, P.:* Nature *191*, 1002 (1961).
48. *Tarte, P.:* Silicates Ind. *28*, 345 (1963).
49. *Tarte, P.:* These d'agregation a l'enseignement supérieur. Universite de Liege, Belgium, 1965.
50. *Bobovich, Ya. S.:* Opt. i Spektroskopiya (USSR) *13*, 459 (1962).
51. *Baran, E. J., Aymonino, P. J.:* Z. Anorg. Allgem. Chem. *365*, 211 (1969).
52. *Baran, E. J., Aymonino, P. J., Müller, A.:* J. Mol. Struct. *11*, 453 (1972).
53. *Baran, E. J.:* Z. anorg. allg. Chem. (in press).
54. *Gopal, R., Calvo, C.:* Z. Krist. *137*, 67 (1973).
55. *Krishnamachari, N., Calvo, C.:* Can. J. Chem. *49*, 1629 (1971).
56. *Tarte, P.:* J. Inorg. Nucl. Chem. *29*, 915 (1967).
57. *Baran, E. J.:* unpublished results.
58. *Zurková, L., Gregorová, M., Dillinger, M.:* Collection Czech. Chem. Commun. *36*, 1906 (1971).
59. *Baran, E. J., Aymonino, P. J.:* Z. Anorg. Allgem. Chem. *383*, 226 (1971).
60. *Baran, E. J., Aymonino, P. J.:* Z. Anorg. Allgem. Chem. *383*, 220 (1971).
61. *Miller, S. A., Caspers, H. H., Rast, H. E.:* Phys. Rev. *168*, 964 (1968).
62. *Baran, E. J., Aymonino, P. J.:* Z. Anorg. Allgem. Chem. *390*, 77 (1972).
63. *Levitt, S. R., Blakeslee, K. C., Condrate Sr., R. A.:* Mem. Soc. Roy. Sci. Liége *20*, 121 (1970).
64. *Levitt, S. R., Condrate Sr., R. A.:* Am. Mineralogist *55*, 1562 (1970).
65. *Klee, W. E., Engel, G.:* J. Inorg. Nucl. Chem. *32*, 1837 (1970).
66. *Banks, E., Greenblatt, M., Schwartz, R. W.:* Inorg. Chem. *7*, 1230 (1968).
67. *Brixner, L. H., Bouchard, R. J.:* Mat. Res. Bull. *5*, 61 (1970).
68. *Blasse, G.:* J. Solid State Chem. *7*, 169 (1973).
69. *Campbell, J. A.:* Spectrochim. Acta *21*, 1333 (1965).
70. *Tarte, P., Nizet, G.:* Spectrochim. Acta *20*, 503 (1964).
71. *Baran, E. J., Aymonino, P. J.:* Anales Asoc. Quim. Arg. *56*, 91 (1968).
72. *Muller, O., White, W. B., Roy, R.:* Spectrochim. Acta *25A*, 1491 (1969).
73. *Cord, P. P., Courtine, P., Pannetier, G., Guillermet, J.:* Spectrochim. Acta *28A*, 1601 (1972).
74. *Darrie, R. G., Doyle, W. P., Kirkpatrick, I.:* J. Inorg. Nucl. Chem. *29*, 979 (1967).
75. *Müller, A., Böschen, I., Baran, E. J.:* Monatsh. Chem. *104*, 821 (1973).
76. *Müller, A., Böschen, I., Baran, E. J., Aymonino, P. J.:* Monatsh. Chem. *104*, 836 (1973).
77. *Müller, A., Baran, E. J., Hendra, P. J.:* Spectrochim. Acta *25A*, 1654 (1969).
78. *Müller, A., Stockburger, M., Baran, E. J.:* Anales Asoc. Quim. Arg. *57*, 65 (1969).
79. *Carter, R. L., Bricker, C. E.:* Spectrochim. Acta *27A*, 569 (1971).
80. *Davies, J. E. D., Long, D. A.:* J. Chem. Soc. A *1971*, 1275.
81. *Hendra, P. J.:* Spectrochim. Acta *24A*, 125 (1968).
82. *Scheuermann, W., Ritter, G. J., Schutte, C. J. H.:* Z. Naturforsch. *25a*, 1856 (1972).
82a. *Carter, R. L.:* Spectr. Letters *5*, 401 (1972).
83. *Baran, E. J., Aymonino, P. J., Müller, A.:* Z. Naturforsch. *24b*, 271 (1969).
84. *Weinstock, N.:* Dissertation, Univ. Dortmund, 1973.
85. *Swanson, H. E., Morris, M. C., Stinchfield, R. P., Evans, N. B.:* Natl. Bur. Stad. (U.S.) Monograph *25*, Sect. I (1962).
86. *Busey, R. H., Keller, O.:* J. Chem. Phys. *41*, 215 (1964).
87. *White, W. B., De Angelis, B. A.:* Spectrochim. Acta *23A*, 985 (1967).
88. *Caillet, P., Saumagne, P.:* J. Mol. Struct. *4*, 191 (1969).
89. *Preudhomme, J., Tarte, P.:* Spectrochim. Acta *28A*, 69 (1972).

90. *Müller, A., Weinstock, N., Mohan, N., Schläpfer, C. W., Nakamoto, K.:* Appl. Spectry. *27*, 257 (1973).
91. *Bues, W., Brockner, W., Grünewald, D.:* Spectrochim. Acta *28A*, 1519 (1972).
92. *Gatehouse, B. M., Leverett, P.:* J. Chem. Soc. *A1969*, 849.
93. *Kools, F. X. N. M., Koster, A. S., Rieck, G. D.:* Acta Cryst. *26B*, 1974 (1970).
94. *Barker, A. S.:* Phys. Rev. *135A*, 742 (1964).
95. *Clark, G. M., Doyle, W. P.:* Spectrochim. Acta *22*, 1441 (1966).
96. *Khanna, R. K., Lippincott, E. R.:* Spectrochim. Acta *24A*, 905 (1968).
97. *Khanna, R. K., Brower, W. S., Guscott, B. R., Lippincott, E. R.:* J. Res. Natl. Bur. Std. *72A*, 81 (1968).
98. *Liegeois-Duyckaerts, M., Tarte, P.:* Spectrochim. Acta *28A*, 2037 (1972).
99. *Porto, S. P. S., Scott, J. F.:* Phys. Rev. *157*, 716 (1967).
100. *Tarte, P., Liegeois-Duyckaerts, M.:* Spectrochim. Acta *28A*, 2029 (1972).
101. *Tarte, P., Preudhomme, J.:* Spectrochim. Acta *26A*, 2207 (1970).
102. *Schwing-Weill, M. J., Arnaud-Neu, F.:* Bull. Soc. Chim. France *1970*, 853.
103. *Brown, R. G., Denning, J., Hallet, A., Ross, S. D.:* Spectrochim. Acta *26A*, 963 (1970).
104. *Lesne, J. P., Cailett, P.:* Can. J. Spectr. *18*, 69 (1973).
105. *Rocchiccioli, C.:* Compt. Rend. *256*, 1707 (1963).
106. *Müller, A., Krebs, B.:* Naturwissenschaften *52*, 448 (1965).
107. *Müller, A., Krebs, B.:* Z. Naturforsch. *21b*, 3 (1966).
108. *Baran, E. J, Aymonino, P. J.:* Monatsh. Chem. *99*, 1584 (1968).
109. *Baran, E. J., Aymonino, P. J.:* Z. Anorg. Allgem. Chem. *354*, 85 (1967).
110. *Baran, E. J.:* Z. Anorg. Allgem. Chem. *382*, 80 (1971).
111. *Doyle, W. P., Kirkpatrick, I.:* Spectrochim. Acta *24A*, 1495 (1968).
112. *Baran, E. J., Aymonino, P. J.:* Monatsh. Chem. *99*, 606 (1968).
113. *Baran, E. J., Aymonino, P. J.:* Z. Anorg. Allgem. Chem. *362*, 215 (1968).
114. *Müller, A., Böschen, I.:* Z. Naturforsch. *26b*, 483 (1971).
115. *Müller, A., Rittner, W.:* Spectrochim. Acta *23A*, 1831 (1967).
116. *Müller, A.:* Z. Naturforsch. *20a*, 745 (1965).
117. *Müller, A.:* Z. Naturforsch. *21a*, 433 (1966).
118. *Ulbricht, K., Kriegsmann, H.:* Z. Chem. *6*, 232 (1966).
119. *Ulbricht, K., Kriegsmann, H.:* Z. Chem. *5*, 276 (1965).
120. *Ulbricht, K., Kriegsmann, H.:* Z. Anorg. Allgem. Chem. *358*, 193 (1968).
121. *Müller, A., Weinstock, N., Baran, E. J.:* Annales Asoc. Quim. Argent. (in press).
121a. *Petrov, K. J., Voronskaja, G. N.:* In: Schwingungsspektren in der Anorganischen Chemie (Russian), (*Y. Y. Kharitonov*, ed.), Moskow: Hayka 1971.
122. *Ueki, T., Zalkin, A., Templeton, D. H.:* Acta Cryst. *19*, 157 (1965).
123. *Levin, I. W.:* Inorg. Chem. *8*, 1018 (1969).
124. *McDowell, R. S.:* Inorg. Chem. *6*, 1759 (1967).
125. *McDowell, R. S., Goldblatt, M.:* Inorg. Chem. *10*, 625 (1971).
126. *McDowell, R. S., Asprey, L. B., Hoskins, L. C.:* J. Chem. Phys. *56*, 5712 (1972).
127. *Griffith, W. P.:* J. Chem. Soc. *1968A*, 1663.
128. *Davidson, G, Logan, N., Morris, A.:* Chem. Commun. *1968*, 1044.
129. *Scholder, R., Klemm, W.:* Angew. Chem. *66*, 461 (1954).
130. *Scholder, R.:* Z. Elektrochem. *56*, 879 (1952).
131. *Scholder, R.:* Angew. Chem. *70*, 583, (1958).
132. *Guerchais, J. E., Leroy, M. J., Rohmer, R.:* Compt. Rend. *261*, 3628 (1965).
133. *Müller, A., Kebabcioglu, R., Leroy, M. J. F., Kaufmann, G.:* Z. Naturforsch. *23b*, 740 (1968).
134. *Doyle, W. P., Eddy, P.:* Spectrochim. Acta *23A*, 1903 (1967).
135. *Baran, E. J., Aymonino, P. J.:* Z. Naturforsch. *23b*, 107 (1968).
136. *Tarte, P., Thelen, J.:* Spectrochim. Acta *28A*, 5 (1972).
137. *Banks, E., Greenblatt, M., McGarvey, B. R.:* J. Chem. Phys. *47*, 3772 (1967).
138. *Baran, E. J.:* Tesis Doctoral; Universidad Nac. de La Plata (1967).
139. *Palenik, G. J.:* Inorg. Chem. *6*, 507 (1967).
140. *Griffith, W. P.:* J. Chem. Soc. *A1966*, 1467.
141. *Campbell, J. A.:* Spectrochim. Acta *21*, 851 (1965).

142. *Audette, R. J., Quail, J. W.:* Inorg. Chem. *11*, 1904 (1972).
143. *Bécarud, N.:* Rapport CEA-R 2895. Centre d'Études Nucleaires de Fontenay Aux Roses (France). 1966.
144. *Porai-Koshits, M. A., Atovmyan, L. O., Andrianov, A. G.:* Zh. Strukt. Khim. USSR *2*, 743 (1961).
145. *Baran, E. J., Müller, A., Kebabcioglu, R., Bollmann, F., Aymonino, P. J.:* Anales Asoc. Quim. Arg. *58*, 247 (1970).
146. *Baran, E. J.:* unpublished data.
147. *Brendel, C., Klemm, W.:* Z. Anorg. Allgem. Chem. *320*, 59 (1963).
148. *Griffith, W. P.:* J. Chem. Soc. *A1969*, 211.
148a. *McCarthy, G. J., White, W. B., Roy, R.:* Inorg. Chem. *8*, 1236 (1969). — *Mumme, W. G., Wadsley, A. D.:* Acta Cryst. *24B*, 1327 (1968). — *Dölling, H., Trömel, M.:* Naturwissenschaften *60*, 153 (1973).
149. *Hauck, J.:* Z. Naturforsch. *25b*, 468 (1970).
150. *Hauck, J.:* Z. Naturforsch. *24b*, 645 (1969).
151. *Hauck, J.:* Z. Naturforsch. *25b*, 224 (1970).
152. *Hauck, J., Fadini, A.:* Z. Naturforsch. *25b*, 422 (1970).
153. *Corsmit, A. F., Hoefdraad, H. E., Blasse, G.:* J. Inorg. Nucl. Chem. *34*, 3401 (1972).
154. *Baran, E. J., Müller, A.:* Z. Anorg. Allgem. Chem. *368*, 168 (1969).
155. *Müller, A., Baran, E. J., Hauck, J.:* Spectrochim. Acta *31A*, 801 (1975).
156. *Müller, A., Weinstock, N., Schulze, H.:* Spectrochim. Acta *28A*, 1075 (1972).
157. *Gattow, G., Franke, A.:* Z. Anorg. Allgem. Chem. *352*, 11 (1967).
158. *Müller, A., Jørgensen, Ch. K., Diemann, E.:* Z. Anorg. Allgem. Chem. *391*, 38 (1972).
159. *Crevecoeur, C.:* Acta Cryst. *17*, 157 (1964).
160. *Müller, A., Sievert, W.:* Z. Anorg. Allgem. Chem. *406*, 80 (1974).
161. *Müller, A., Schmidt, K. H., Tytko, K. H., Bouwma, J., Jellinek, F.:* Spectrochim. Acta *28A*, 381 (1972).
162. *Müller, A., Diemann, E., Rao, V. V. K.:* Chem. Ber. *103*, 2961 (1970).
163. *Müller, A., Krebs, B., Kebabcioglu, R., Stockburger, M., Glemser, O.:* Spectrochim. Acta *24A*, 1831 (1968).
164. *Carter, R. L., Bricker, C. E.:* Spectrochim. Acta *27A*, 825 (1971).
165. *Müller, A., et al.:* in preparation.
166. *Müller, A., Diemann, E., Heidborn, U.:* Z. Anorg. Allgem. Chem. *376*, 125 (1970).
167. *Müller, A., Sievert, W., Schulze, H., Weinstock, N.:* Z. Anorg. Allgem. Chem. *403*, 310 (1974).
168. *Müller, A., Weinstock, N., Krebs, B., Buss, B., Ferwanah, A.:* Z. Naturforsch. *26b*, 268 (1971).
169. *Müller, A., Sievert, W., Schulze, H.:* Z. Naturforsch. *27b*, 720 (1972).
169a. *Müller, A., Schulze, H., Sievert, W., Weinstock, N.:* Z. Anorg. Allgem. Chem. *403*, 310 (1974).
170. *Schmidt, K. H., Müller, A.:* Spectrochim. Acta *28A*, 1829 (1972).
171. *Müller, A., Diemann, E., Baran, E. J.:* Z. Anorg. Allgem. Chem. *375*, 87 (1970).
172. *Gonschorek, W., Hahn, Th., Müller, A.:* Z. Krist. *138*, 380 (1973).
173. *Schmidt, K. H., Müller, A., Bouwma, J., Jellinek, F.:* J. Mol. Struct. *11*, 275 (1972).
174. *Müller, A., Sievert, W.:* Z. Naturforsch. *27b*, 722 (1972).
175. *Baran, E. J., Aymonino, P. J.:* Z. Naturforsch. *27b*, 76 (1972).
176. *Müller, A., Weinstock, N., Mohan, N., Schlaepfer, C. W., Nakamoto, K.:* Z. Naturforsch. *27a*, 542 (1972).
177. *Preudhomme, J., Tarte, P.:* Spectrochim. Acta *27A*, 1817 (1971).
178. *Tarte, P., Preudhomme, J.:* Spectrochim. Acta *29A*, 1301 (1973).
179. *Preudhomme, J.:* Thése, Liége, 1970.
180. *Guerchais, J. E., Haumesser, A., Rohmer, R.:* Compt. Rend. *260*, 5571 (1965).
181. *Adler, H. H., Kerr, P. F.:* Am. Mineralogist *48*, 124 (1963).
182. *Adler, H. H., Kerr, P. F.:* Am. Mineralogist *48*, 839 (1963).
183. *Adler, H. H., Kerr, P. F.:* Am. Mineralogist *50*, 132 (1965).
184. *Weir, C. E., Lippincott, E. R.:* J. Res. Natl. Bur. Std. *65A*, 173 (1961).
185. *Adler, H. H.:* Am. Mineralogist *50*, 1553 (1965).

186. *Laperches, J. P., Tarte, P.:* Spectrochim. Acta *22*, 1201 (1966).
187. *Baran, E. J., Aymonino, P. J.:* Anales Asoc. Quim. Arg. *56*, 11 (1968).
188. *Müller, A., Krebs, B., Gattow, G.:* Spectrochim. Acta *23A*, 2809 (1967).
189. *Müller, A., Diemann, E., Krebs, B., Leroy, M. J. F.:* Angew. Chem. *80*, 846 (1968).
190. *Krebs, B., Müller, A., Gattow, G.:* Z. Naturforsch. *20b*, 1017 (1965).
191. *Müller, A., Baran, E. J.:* J. Mol. Struct. *15*, 283 (1973).
192. *Müller, A., Krebs, B.:* J. Mol. Spectry. *24*, 180 (1967).
193. *Chaves, A., Porto, S. P. S.:* Solid State Commun. *10*, 1075 (1972).
194. *Adams, D. M., Hooper, M. A., Lloyd, M. H.:* J. Chem. Soc. *A 1971*, 946.
194a. *Carter, R. O., Königer-Ahlborn, E., Müller, A.:* to be submitted.
195. *Scott, J. F.:* J. Chem. Phys. *48*, 874 (1968).
196. *Scott, J. F.:* J. Chem. Phys. *49*, 98 (1968).
197. *Brown, R. S., Denning, J., Hallet, H., Ross, S. D.:* Spectrochim. Acta *26A*, 963 (1970).
198. *Nicol, M., Durana, J. F.:* J. Chem. Phys. *54*, 1436 (1971).
199. *Miller, P. J.:* Spectrochim. Acta *27A*, 957 (1971).
200. *Johnson, R. A., Rogers, M. T., Leroi, G. E.:* J. Chem. Phys. *56*, 789 (1972).
201. *Becher, H. J., Friedrich, F., Willner, H.:* Z. Anorg. Allgem. Chem. *395*, 134 (1973).
202. *Sherwood, P. M. A.:* Spectrochim. Acta *27A*, 1019 (1971).
203. *Baran, E. J., Müller, A.:* Spectrochim. Acta *27A*, 517 (1971).
204. *Price, W. C., Sherman, W. F., Wilkinson, G. R.:* Spectrochim. Acta *16*, 663 (1960).
205. *Decius, J. C., Coker, E. H., Brenna, G. L.:* Spectrochim. Acta *19*, 1281 (1963).
206. *Krynaw, G. N., Schutte, C. J. H.:* Spectrochim. Acta *21*, 1947 (1965).
207. *Kato, R., Rolfe, J.:* J. Chem. Phys. *47*, 1901 (1967).
208. *Klee, W. E.:* Z. Anorg. Allgem. Chem. *370*, 1 (1969).
209. *Klee, W. E.:* Spectrochim. Acta *26A*, 1165 (1970).
210. *Tarte, P.:* Spectrochim. Acta *18*, 467 (1962).
211. *Tarte, P.:* Spectrochim. Acta *19*, 25 (1963).
212. *Tarte, P.:* Spectrochim. Acta *19*, 49 (1963).
213. *Steger, E., Danzer, K.:* Ber. Bunsenges. Physik. Chem. *68*, 635 (1964).
214. *Barraclough, C. G., Bilander, I.:* Australian J. Chem. *23*, 1471 (1970).
215. *Müller, A., Königer, F., Weinstock, N.:* Spectrochim. Acta *30A*, 641 (1974).
216. *Miller, P. J., Cessac, G. L., Khanna, R. K.:* Spectrochim. Acta *27A*, 2019 (1971).
217. *Ferraro, J. R.:* In: Spectroscopy in Inorganic chemistry, Vol. II (*C. N. R. Rao* and *J. R. Ferraro*, eds.). New York: Academic Press 1971.
218. *Whyman, R.:* In: Laboratory Methods in IR-spectroscopy (*R. G. J. Miller* and *B. C. Stace*, eds.). London: Heyden 1972.
219. *Adams, D. M., Payne, S. J.:* Annual Reports, Sect. A, The Chem. Soc., 1972.
220. *Nakamoto, K., Udovich, C., Ferraro, J. R., Quattrochi, A.:* Appl. Spectry. *24*, 606 (1970).
221. *Schutte, C. J. H., Heyns, A. M.:* J. Mol. Struct. *5*, 37 (1970).
222. *Wilks, P. A.:* In: Laboratory methods in IR-spectroscopy (*R. G. J. Miller* and *B. C. Stace*, eds.). London: Heyden 1972.
223. *Brooker, M. H.:* J. Chem. Phys. *53*, 4100 (1970).
224. *D'Ambrogio, F., Brüesch, P., Kalbfleisch, H.:* Phys. Status Solidi (B) *49*, 117 (1972).
225. *Scheuermann, W., Schutte, C. J. H.:* J. Raman Spectr. *1*, 605 (1973).
226. *Kiefer, W., Bernstein, H. J.:* Chem. Phys. Letters *8*, 381 (1971).
227. *Kiefer, W., Bernstein, H. J.:* Appl. Spectry. *25*, 609 (1971).
228. *Kiefer, W., Bernstein, H. J.:* Mol. Phys. *23*, 835 (1972).
229. *Kiefer, W., Müller, A.:* unpublished data.
230. *Ranade, A., Stockburger, M.:* Chem. Phys. Letters *22*, 257 (1973).
231. *Ranade, A., Krasser, W., Müller, A., Ahlborn, E.:* Spectrochim. Acta *30A*, 1341 (1974).
232. *Müller, A., Ahlborn, E.:* Spectrochim. Acta *31A*, 75 (1975).
233. *Müller, A., Diemann, E.:* J. Chem. Phys. *61*, 5469 (1974).
234. *Derkosch, J.:* Absorptionsspektralanalyse im UV, Sichtbaren und IR-Gebiet. Frankfurt/M.: Akad. Verlagsges. 1967.
235. *Miller, R. G. J., Stace, B. C.*, eds.: Laboratory methods in IR-spectroscopy. London: Heyden 1972.
236. *Hill, D. G., Rosenberg, A. F.:* J. Chem. Phys. *24*, 1219 (1956).

237. *Meloche, V. W., Kalbus, G. E.:* J. Inorg. Nucl. Chem. *6*, 104 (1958).
238. *Baran, E. J., Aymonino, P. J.:* Spectrochim. Acta *24A*, 288 (1968).
239. *Castinel, C., Dupuy, B., Garrigou-Lagrange, Ch.:* Bull. Soc. Chim. France *1969*, 1394.
240. *Lauwers, H., Desseyn, H.:* J. Inorg. Nucl. Chem. *36*, 475 (1974).
241. *McDevitt, N. T., Baun, W. L.:* Spectrochim. Acta *20*, 799 (1964).
242. *Pedregosa, J., Baran, E. J., Aymonino, P. J.:* Z. Anorg. Allgem. Chem. *404*, 308 (1974).
243. *Kutzelnigg, W., Nonnenmacher, G., Mecke, R.:* Chem. Ber. *93*, 1279 (1960).
244. *Schroeder, R. A., Lippincott, E. R., Weir, C. E.:* J. Inorg. Nucl. Chem. *28*, 1397 (1966).
245. *Baran, E. J., Gentil, L. A., Pedregosa, J. C., Aymonino, P. J.:* Z. Anorg. Allgem. Chem. *410*, 301 (1974).

Received May 9, 1975

Author-Index Volume 1-26

S. P. SINHA

Europium

23 figures. VIII, 164 pages. 1967
(Anorganische und allgemeine Chemie
in Einzeldarstellungen, Band 8)
This volume provides the first unified
treatment of the coordination chemistry and spectral characteristics of
europium and its complexes. The
subject has been reviewed balancing
the experimental facts with the theory
as far as possible.

K. D. GUNDERMANN

**Chemilumineszenz organischer
Verbindungen**

33 Abbildungen. VII, 174 Seiten. 1968
(Organische Chemie in Einzeldarstellungen, Band 11)

Aus den Besprechungen:
„Die Entwicklung des im vorliegenden
Band beschriebenen Forschungsgebie-
tes hat nach dem Zweiten Weltkrieg
stürmische Formen angenommen, was
vor allem durch die immer empfind-
licheren photoelektrischen Meßgeräte
und neue physikalisch-chemische Ar-
beitsmethoden ausgelöst wurde. Die
Fülle des Materials rechtfertigt es, die
Ergebnisse aus der Sicht des Orga-
nikers in einer gesonderten Mono-
graphie darzustellen.
Die Präsentation des Stoffes erfolgt in
einer Weise, daß sich auch der Nicht-
Fachmann einarbeiten kann, während
der Spezialist viele Anregungen für
weitere Untersuchungen zu gewinnen
vermag."
Chemische Rundschau

A. SCHÖNBERG

**Preparative Organic
Photochemistry**

In cooperation with G. O. Schenk,
O. A. Neumüller. Second completely
revised edition of Präparative Organische Photochemie. 4 figures.
51 tables. XXIV, 608 pages. 1968
This monograph is an exhaustive compilation of photochemical reactions
that are of interest to the organic
chemist. Sufficient experimental details are given to make this book a
highly useful manual of photochemical
laboratory techniques for the student
and the specialist.

**Springer-Verlag
Berlin Heidelberg New York**

CH. K. JØRGENSEN

Oxidation Numbers and Oxidation States

VII, 291 pages. 1969
(Molekülverbindungen und Koordinationsverbin-
dungen in Einzeldarstellungen)

Contents: Formal Oxidation Numbers. Configura-
tions in Atomic Spectroscopy. Characteristics of
Transition Group Ions. Internal Transitions in
Partly Filled Shells. Inter-Shell Transitions. Electron
Transfer Spectra and Collectively Oxidized Ligands.
Oxidation States in Metals and Black Semi-Con-
ductors. Closed-Shell Systems, Hydrides and Back-
Bonding. Homopolar Bonds and Catenation.
Quanticule Oxidation States. Taxological Quantum
Chemistry.

H.-H. PERKAMPUS

**Wechselwirkung von Pi-Elektronensystemen
mit Metallhalogeniden**

64 Abbildungen. XI, 215 Seiten. 1973
(Molekülverbindungen und Koordinationsverbin-
dungen in Einzeldarstellungen)

Inhaltsübersicht: Einleitung und Abgrenzung. Eigen-
schaften der Donatoren und Acceptoren. Proton-
Additions-Komplexe. π-Komplexe. σ-Komplexe.

G. KORTÜM

Reflectance Spectroscopy

Principles, Methods, Applications.
Translator from the German: J. E. Lohr.
160 figures. VI, 366 pages. 1969

Contents: Introduction. Regular and Diffuse
Reflection. Single and Multiple Scattering. Phenomenological Theories of Absorption and Scattering
of Tightly Packed Particles. Experimental Testing
of the "Kubelka-Munk" Theory. Experimental Techniques. Applications. Reflectance Spectra Obtained
by Attenuated Total Reflection. Appendices.
Subject Index.

Springer-Verlag
Berlin
Heidelberg
New York